The IMA Volumes in Mathematics and Its Applications

Volume 9

Series Editors
George R. Sell   Hans Weinberger

Institute for Mathematics and Its Applications
IMA

The **Institute for Mathematics and Its Applications** was established by a grant from the National Science Foundation to the University of Minnesota in 1982. The IMA seeks to encourage the development and study of fresh mathematical concepts and questions of concern to the other sciences by bringing together mathematicians and scientists from diverse fields in an atmosphere that will stimulate discussion and collaboration.

The IMA Volumes are intended to involve the broader scientific community in this process.

Hans Weinberger, Director
George R. Sell, Associate Director

IMA Programs

1982–1983 Statistical and Continuum Approaches to Phase Transition

1983–1984 Mathematical Models for the Economics of Decentralized Resource Allocation

1984–1985 Continuum Physics and Partial Differential Equations

1985–1986 Stochastic Differential Equations and Their Applications

1986–1987 Scientific Computation

1987–1988 Applied Combinatorics

1988–1989 Nonlinear Waves

Springer Lecture Notes from the IMA

The Mathematics and Physics of Disordered Media
Editors: Barry Hughes and Barry Ninham
(Lecture Notes in Mathematics, Volume 1035, 1983)

Orienting Polymers
Editor: J. L. Ericksen
(Lecture Notes in Mathematics, Volume 1063, 1984)

New Perspectives in Thermodynamics
Editor: James Serrin
(Springer-Verlag, 1986)

Models of Econoic Dynamics
Editor: Hugo Sonnenschein
(Lecture Notes In Economics, Volume 264, 1986)

George Papanicolaou
Editor

Hydrodynamic Behavior and Interacting Particle Systems

With 10 Illustrations

Springer-Verlag
New York Berlin Heidelberg
London Paris Tokyo

George Papanicolaou
Courant Institute for
Mathematical Sciences
New York University
New York, New York 10012

AMS Classification: 60K35 76T05 82A42 82A70

Library of Congress Cataloging in Publication Data
Hydrodynamic behavior and interacting particle
systems.
(The IMA volumes in mathematics and its
applications ; v. 9)
Papers presented at a workshop, held at the
Institute for Mathematics and its Applications of
the University of Minnesota during the week of
March 17, 1986.
1. Two-phase flow—Congresses. 2. Multiphase
flow—Congresses. 3. Stochastic processes—
Congresses. 4. Transport theory—Congresses.
I. Papanicolaou, George. II. University of Minnesota.
Institute for Mathematics and its Applications.
III. Series.
QA922.H93 1987 532′.051 87-16700

Text prepared by the author in camera-ready form.
Printed and bound by Edwards Brothers Inc., Ann Arbor, Michigan.
Printed in the United States of America.

9 8 7 6 5 4 3 2 1

ISBN 0-387-96584-X Springer-Verlag New York Berlin Heidelberg
ISBN 3-540-96584-X Springer-Verlag Berlin Heidelberg New York

The IMA Volumes in Mathematics and Its Applications

Current Volumes:

Volume 1: Homogenization and Effective Moduli of Materials and Media
Editors: Jerry Ericksen, David Kinderlehrer, Robert Kohn, and J.-L. Lions

Volume 2: Oscillation Theory, Computation, and Methods of Compensated Compactness
Editors: Constantine Dafermos, Jerry Ericksen, David Kinderlehrer, and Marshall Slemrod

Volume 3: Metastability and Incompletely Posed Problems
Editors: Stuart Antman, Jerry Ericksen, David Kinderlehrer, and Ingo Müller

Volume 4: Dynamical Problems in Continuum Physics
Editors: Jerry Bona, Constantine Dafermos, Jerry Ericksen, and David Kinderlehrer

Volume 5: Theory and Application of Liquid Crystals
Editors: Jerry Erickson and David Kinderlehrer

Volume 6: Amorphous Polymers and Non-Newtonian Fluids
Editors: Constantine Dafermos, Jerry Ericksen, and David Kinderlehrer

Volume 7: Random Media
Editor: George Papanicolaou

Volume 8: Percolation Theory and Ergodic Theory of Infinite Particle Systems
Editor: Harry Kesten

Volume 9: Hydrodynamic Behavior and Interacting Particle Systems
Editor: George Papanicolaou

Forthcoming Volumes:

1985–1986: Stochastic Differential Equations and Their Applications
Stochastic Differential Systems, Stochastic Control Theory and Applications

1986–1987: Scientific Computation
Computational Fluid Dynamics and Reacting Gas Flows
Numerical Algorithms for Modern Parallel Computer Architectures
Numerical Simulation in Oil Recovery
Atomic and Molecular Structure and Dynamics

CONTENTS

FOREWORD

This IMA Volume in Mathematics and its Applications

HYDRODYNAMIC BEHAVIOR
AND INTERACTING PARTICLE SYSTEMS

is in part the proceedings of a workshop which was an integral part of the 1985-86 IMA program on STOCHASTIC DIFFERENTIAL EQUATIONS AND THEIR APPLICATIONS. We are grateful to the Scientific Committee:

Daniel Stroock (Chairman)
Wendell Fleming
Theodore Harris
Pierre-Louis Lions
Steven Orey
George Papanicolaou

for planning and implementing an exciting and stimulating year-long program. We especially thank the Program Organizer, George Papanicolaou for organizing a workshop which brought together scientists and mathematicians in a variety of areas for a fruitful exchange of ideas.

George R. Sell

Hans Weinberger

PREFACE

A workshop on the hydrodynamic behavior of interacting particle systems was held at the Institute for Mathematics and its Applications at the University of Minnesota during the week of March 17, 1986. Fifteen papers presented at the workshop are collected in this volume. They contain research in several different directions that are currently being pursued.

The paper of Chaikin, Dozier and Lindsay is concerned with experimental results on suspensions in regimes where modern mathematical methods could be useful. The paper of Fritz gives an introduction to these methods as does the paper of Spohn. Analytical methods currently used by in the physics and chemistry literature are presented in the paper of Freed, Wang and Douglas. The paper of Caflisch deals with time dependent effects in sedimentation.

In the papers by Ozawa, Rubinstein and Figari, Papanicolaou, Rubinstein the continuum limit of boundary value problems in regions with many small inclusions is analyzed. These are static problems but one expects that the methods used will be useful eventually for dynamic problems where the inclusions (particles) move.

The vortex method as a problem in numerical analysis is treated by Goodman and as a probabilistic one by Osada. The propagation of chaos for Burgers equation is treated by Sznitman. The papers of Dawson, Gorostiza, and Tanaka deal with probabilistic aspects of particle systems (or random media) while Loper's paper introduces a continuum mechanics model for flow of a slurry.

George Papanicolaou

STOCHASTIC MODELLING OF A DILUTE FLUID-PARTICLE SUSPENSION *

Russel E. Caflisch **

Courant Institute of Mathematical Sciences
New York University

Abstract

The bulk properties (such as average sedimentation speed and diffusion coefficient) in a sedimenting suspension of solid particles are sensitive to the statistical distribution of the particles. In this paper a method for determining the two particle distribution is described. In this method the particles are treated as point forces and only three particle interactions are included. Under several approximations it is shown that the two particle distribution separates into a product of functions of distance between the pair and of angle between the pair and the vertical.

---

* Presented at the Workshop on Hydrodynamic Behavior and Interacting Particle Systems and Applications, Institute for Mathematics and its Applications, University of Minnesota, Minneapolis

** Research partially supported by the National Science Foundation under contract No. NSF-DMS-83-12229

I. Introduction

Consider a suspension of solid spheres in viscous fluid. The spheres are sedimenting through the fluid due to gravity. We neglect inter-particle forces between the spheres, except hydrodynamic forces, i.e. the motion of one sphere sets the fluid in motion and through fluid pressure and viscosity exerts a force on a second particle. Throughout most of this paper, Brownian motion of the particles is ignored.

The goal here is determination of the statistical equilibrium state for such a suspension. The suspension is assumed to be spatially homogenous, at least on a fairly large scale. The effect of the bottom is neglected, i.e. the desired equilibrium is not the rest state in which all particles sit on the bottom, but a moving quasi-equilibrium.

The motivation for investigating the particle distribution is that bulk properties are sensitive to the particle statistics. These bulk properties include average sedimentation speed, coefficient of (thermal) Brownian motion and coefficient of hydrodynamic dispersion. By hydrodynamic dispersion we mean the random motion of sedimenting particle that is caused by its interactions with the other, randomly distributed, sedimenting particles.

II. Stokes Equations for a Suspension

In a sedimenting suspension, inertial forces are negligible if the Reynolds numer is small. The appropriate Reynolds number is that based on particle size, density and velocity, i.e.

$$Re = \frac{\rho_p \, a|\underset{\sim}{V}_{st}|}{\mu} \ll 1 \tag{2.1}$$

in which

ρ_p = particle mass density

a = particle radius

μ = fluid viscosity

$\underset{\sim}{V}_{st}$ = the characteristic particle velocity

$$= \frac{2a^2}{9\pi\mu} \underset{\sim}{g} \ . \tag{2.2}$$

Let the particles have centers $\underset{\sim}{x}_1, \ldots, \underset{\sim}{x}_N$ and velocities $\underset{\sim}{V}_1, \ldots, \underset{\sim}{V}_N$. Stokes equations for the fluid velocity $\underset{\sim}{u}$ and pressure p are

$$\mu \Delta^2 \underset{\sim}{u} - \underset{\sim}{\nabla} p = - \text{Force}$$
$$\underset{\sim}{\nabla} \cdot \underset{\sim}{u} = 0 , \qquad (2.3)$$

in which the force is concentrated on the particle boundaries $|\underset{\sim}{x} - \underset{\sim}{x}_i| = a$. A more explicit description of the Stokes equation is given in [3,4]. In this paper we shall employ the point-particle approximation.

The average sedimentation speed is defined as the statistical average of the $\underset{\sim}{V}_i$'s (assuming the particles are treated as indistinguishable) i.e.

$$\underset{\sim}{V}_{sed} = \langle \underset{\sim}{V}_1 \rangle . \qquad (2.4)$$

The value $\underset{\sim}{V}_{sed}$ has been determined under various assumptions about the particle distribution. If the particles are distributed periodically (hence deterministically) with lattice spacing ℓ, Hasimoto [5] showed that

$$\underset{\sim}{V}_{sed} = \underset{\sim}{V}_{st} \left(1 - c\, \phi^{1/3}\right) \qquad (2.5)$$

in which the constant c depends on the type of lattice. If the particles are distributed uniformly and independently except that they cannot overlap (the hard-spheres distribution, Batchelor [1] showed that

$$\underset{\sim}{V}_{sed} = \underset{\sim}{V}_{st} (1 - 6.55\, \phi) . \qquad (2.6)$$

In equation (2.5), (2.6) ϕ is the volume fraction, given by

$$\phi = n \left(\frac{4}{3} \pi a^3\right) = N \left(\frac{4}{3} \pi a^3\right) / |\Omega| \qquad (2.7)$$

in which n is the particle number density and $|\Omega|$ is the volume of the region containing the N particles. Also the average interparticle spacing is

$$\ell = n^{-1/3} = (|\Omega| / N)^{1/3} \tag{2.8}$$

The difference between the exponents 1 and 1/3 shows the sensitivity of $\underset{\sim}{V}_{sed}$ to particle statistics. A unified theory for (2.5), (2.6) was begun by Saffman [7] and developed by Rubinstein [6]. In particular Rubinstein showed how exponents between 1/3 and 1 could be obtained.

The variance in the particle speed $\underset{\sim}{V}$ was calculated by Caflisch and Luke [2] for the same hard-spheres distribution used by Batchelor [1]. They showed that the calculated variance is unphysically large, which seems to cast doubt on the validity of the hard-spheres distribution.

III. Point Particle Model

If the particles are very small compared to their inter-particle spacing, i.e. $a << \ell$, they may be treated as points. The force exerted by a particle on the fluid is the same as the total external force on the particle, which is $\underset{\sim}{F} = (\frac{4}{3}\rho_p a^3)\underset{\sim}{g} = 6\pi\mu a\, \underset{\sim}{v}_{st}$. We scale the problem so that as $a \to 0$, $\underset{\sim}{F}$ remains constant. Thus each particle acts as a point force in the fluid. Stokes equations for a suspension of point particles are then

$$\begin{aligned} \mu\nabla^2 \underset{\sim}{u} - \underset{\sim}{\nabla} p &= \underset{\sim}{F} \sum_{i=1}^{N} \delta(\underset{\sim}{x} - \underset{\sim}{x}_i) \\ \underset{\sim}{\nabla} \cdot \underset{\sim}{u} &= 0 \end{aligned} \tag{3.1}$$

which has solution

$$\underset{\sim}{u}(\underset{\sim}{x}) = |\underset{\sim}{F}| \sum_{i=1}^{N} \underset{\sim}{U}(\underset{\sim}{x} - \underset{\sim}{x}_i) \tag{3.2}$$

in which $\underset{\sim}{U}$ is the stokeslet given by

$$\underset{\sim}{U}(\underset{\sim}{x}) = \frac{\underset{\sim}{e}}{|x|} + \frac{\underset{\sim}{e} \cdot x}{|x|^3}\, \underset{\sim}{x} \tag{3.3}$$

and $\underset{\sim}{e} = \underset{\sim}{F} / |\underset{\sim}{F}|$.

According to Faxen's laws [3,4] the velocity of the ith particle in such a flow is (for $a = 0$)

$$\dot{\underset{\sim}{x}}_i = \underset{\sim}{v}_i = \underset{\sim}{v}_{st} + |\underset{\sim}{F}| \sum_{j \neq i} \underset{\sim}{U}(\underset{\sim}{x}_i - \underset{\sim}{x}_j) \tag{3.4}$$

The system of equations (3.4) describe the motion of a suspension of point particles. Simplify equation (3.4) by a galilean transformation to remove the term $\underset{\sim}{v}_{st}$ and by setting $|F| = 1$, to get

$$\dot{x}_i = \sum_{j \neq i} \underset{\sim}{U}(x_i - x_j). \tag{3.4'}$$

If there are only two particles, $N = 2$, then

$$\underset{\sim}{v}_1 = \dot{\underset{\sim}{x}}_1 = \underset{\sim}{U}(\underset{\sim}{x}_1 - \underset{\sim}{x}_2)$$
$$\underset{\sim}{v}_2 = \dot{\underset{\sim}{x}}_2 = \underset{\sim}{U}(\underset{\sim}{x}_2 - \underset{\sim}{x}_1). \tag{3.5}$$

Since $\underset{\sim}{U}$ is even, then $\underset{\sim}{v}_1 = \underset{\sim}{v}_2$, that is two isolated particles fall together without changing their separation. Thus if only two particle interactions are included, any two-particle distribution function is an equilibrium.

IV. Three Particle Problem

The equations for three particles are

$$\dot{\underset{\sim}{x}}_1 = \underset{\sim}{U}(\underset{\sim}{x}_1 - \underset{\sim}{x}_2) + \underset{\sim}{U}(\underset{\sim}{x}_1 - \underset{\sim}{x}_3)$$
$$\dot{\underset{\sim}{x}}_2 = \underset{\sim}{U}(\underset{\sim}{x}_2 - \underset{\sim}{x}_1) + \underset{\sim}{U}(\underset{\sim}{x}_2 - \underset{\sim}{x}_3) \tag{4.1}$$
$$\dot{\underset{\sim}{x}}_3 = U(\underset{\sim}{x}_3 - \underset{\sim}{x}_1) + \underset{\sim}{U}(\underset{\sim}{x}_3 - \underset{\sim}{x}_2)$$

If the three particles are located on a horizontal equilateral triangle, then they fall together without changing their relative positions. However this equilibrium configuration can be shown to be unstable, although only algebraically so.

For any other 3-particle configuration, we belive that the particle trajectories can be characterized as follows: at t near $-\infty$ there is a pair of particles very far above the third single particle. As time progresses the pair falls toward the single particle (since two particles fall faster than one). As they get closer the three particles interact, but eventually a pair of them (not necessarily the original pair) will pull away from the third. At t near $+\infty$, the pair will be very far below the single particle. A particle pair may be described by the distance r between the pair, and the cosine α of the angle that the pair makes with the vertical. Thus the three particle interaction may be thought of as a scattering event, by which an initial pair described by (r_i, α_i) (near $t = -\infty$) goes to a final pair described by (r_f, α_f) (near $t = +\infty$). In this collision between a pair and a single particle, the relative position of the single particle can be described by impact parameters, just as in the kinetic theory of gases.

In order to show that the scattering scenario described above makes sense, we must show that the pair separation coordinates $(r, \alpha)(t)$ have finite limits at $t = \pm\infty$ and that impact parameters can be defined. For this purpose, equation (4.1) are solved asymptotically for a pair (say $\underset{\sim}{x}_1, \underset{\sim}{x}_2$) far separated from a third particle. Denote

$$\underset{\sim}{z} = \frac{1}{2}(x_1 - x_2) \quad , \quad \underset{\sim}{y} = \frac{1}{2}(\underset{\sim}{x}_1 + \underset{\sim}{x}_2) - \underset{\sim}{x}_3.$$

Then (4.1) is equivalent to

$$\dot{\underset{\sim}{z}} = \frac{1}{2}\underset{\sim}{U}(\underset{\sim}{y} + \underset{\sim}{z}) - \frac{1}{2}\underset{\sim}{U}(\underset{\sim}{y} - \underset{\sim}{z}) \tag{4.2}$$

$$\dot{\underset{\sim}{y}} = \underset{\sim}{U}(2\underset{\sim}{z}) - \frac{1}{2}\underset{\sim}{U}(\underset{\sim}{y} + \underset{\sim}{z}) - \frac{1}{2}\underset{\sim}{U}(\underset{\sim}{y} - \underset{\sim}{z}) \tag{4.3}$$

Assume that

$$|\underset{\sim}{z}| \ll |\underset{\sim}{y}|. \tag{4.4}$$

Since $\underset{\sim}{U}(\underset{\sim}{x}) = O(|\underset{\sim}{x}|^{-1})$, equation (4.2),(4.3) become

$$\dot{\underset{\sim}{z}} = (\underset{\sim}{z} \cdot \underset{\sim}{\nabla}) \underset{\sim}{U} (\underset{\sim}{y}) + O(|\underset{\sim}{z}|^2 / |\underset{\sim}{y}|^3) \qquad (4.5)$$

$$\dot{\underset{\sim}{y}} = \underset{\sim}{U}(2\underset{\sim}{z}) - \underset{\sim}{U}(\underset{\sim}{y}) + O(|\underset{\sim}{z}| / |\underset{\sim}{y}|^2) \qquad (4.6)$$

Expand z, y as

$$z = z_o + z_1 + z_2 + \ldots \qquad (4.7)$$

$$y = y_o + y_1 + y_2 + \ldots$$

Then

$$\dot{\underset{\sim}{z}}_o = 0 \qquad (4.8)$$

$$\dot{\underset{\sim}{y}}_o = \underset{\sim}{v}_o = \underset{\sim}{U}(2\underset{\sim}{z}_o) \qquad (4.9)$$

$$\dot{\underset{\sim}{z}}_1 = (\underset{\sim}{z}_o \cdot \underset{\sim}{\nabla}) \underset{\sim}{U} (\underset{\sim}{y}_o) \qquad (4.10)$$

$$\dot{\underset{\sim}{y}}_1 = 2(\underset{\sim}{z}_1 \cdot \underset{\sim}{\nabla}) \underset{\sim}{U} (2\underset{\sim}{z}_o) - \underset{\sim}{U} (\underset{\sim}{y}_o) \qquad (4.11)$$

Equation (4.8), (4.9) have solutions

$$\underset{\sim}{z}_o(t) = \underset{\sim}{z}_o \qquad (4.12)$$

$$\underset{\sim}{y}_o(t) = \underset{\sim}{v}_o (t - t_o) + \underset{\sim}{y}_\perp \qquad (4.13)$$

in which $\underset{\sim}{z}_o$, $\underset{\sim}{y}_\perp$, t_o are arbitrary constants with $\underset{\sim}{y}_\perp \cdot \underset{\sim}{v}_o = 0$. For $|t-t_o|$ large (4.10) is approximately

$$\dot{\underset{\sim}{z}}_1 = (\operatorname{sgn}(t-t_o)) (t-t_o)^{-2} \underset{\sim}{u}_1 \qquad (4.14)$$

in which $\underset{\sim}{u}_1$ is a constant vector

$$\underset{\sim}{u}_1 = (\underset{\sim}{z}_o \cdot \underset{\sim}{\nabla}) U (\underset{\sim}{v}_o). \qquad (4.15)$$

This has solution

$$\underset{\sim}{z}_1(t) = -|t - t_o|^{-1} \underset{\sim}{u}_1 \qquad (4.16)$$

The equation (4.11) for $\underset{\sim}{y}_1$ is approximately

$$\dot{\underset{\sim}{y}}_1 = |t-t_o|^{-1} \underset{\sim}{v}_1 \qquad (4.17)$$

with

$$\underset{\sim}{v}_1 = -\frac{1}{2} (\underset{\sim}{u}_1 \cdot \nabla) \underset{\sim}{U} (\underset{\sim}{z}_o) - \underset{\sim}{U}(\underset{\sim}{v}_o). \qquad (4.18)$$

It follows that

$$\underset{\sim}{y}_1(t) = \mathrm{sgn}\,(t-t_o) \log |1-t/t_o| \underset{\sim}{v}_1 . \qquad (4.19)$$

Possible constants are omitted from $\underset{\sim}{y}_1$ and $\underset{\sim}{z}_1$ since they are included in $\underset{\sim}{y}_o$, $\underset{\sim}{z}_o$.

In summary the expansions (4.7) of z, y for large time are

$$\underset{\sim}{z}(t) = \underset{\sim}{z}_o - |t-t_o|^{-1} \underset{\sim}{u}_1 + 0(|t-t_o|^{-2} \log |t-t_o|) \qquad (4.20)$$

$$\underset{\sim}{y}(t) = (t-t_o) \underset{\sim}{v}_o + \underset{\sim}{y}_\perp + \mathrm{sgn}\,(t-t_o) \log |1-t/t_o| \underset{\sim}{v}_1$$

$$+ 0(|t-t_o|^{-1} \log |t-t_o|) \qquad (4.21)$$

Note that the expansion is only valid if $|t| >> 1$. The constants $\underset{\sim}{z}_o, t_o, \underset{\sim}{y}_\perp$ for t near ∞ may differ from those for t near $-\infty$. This shows that the pair separation $\underset{\sim}{z}$ has a limit $\underset{\sim}{z}$ as $t \to \infty$ or as $t \to \infty$ (with a different $\underset{\sim}{z}_o$).

We define $\underset{\sim}{y}_\perp$ as the impact coordinates for the single particle in a plane perpendicular to the relative velocity $\underset{\sim}{v}_o$ between the pair and single particle. Note that there is a logarithmic correction to the impact parameter, but that this correction $\underset{\sim}{v}_1$ (defined by (4.18)), as well as $\underset{\sim}{v}_o$ is independent of $\underset{\sim}{y}_\perp$. This allows us to assume that in a randomly chosen collision the impact corrdinates $\underset{\sim}{y}_\perp$ are independent from the pair coordinates $r = 2\,|\underset{\sim}{z}_o|$, $\alpha = (\underset{\sim}{e} \cdot \underset{\sim}{z}_o \,/\, |\underset{\sim}{z}_o|)$. Moreover, the effect of a collision is described by a scattering mapping S from (r_i, α_i) to (r_f, α_f) depending $\underset{\sim}{y}_\perp$, i.e.

$$r_f = R\,(\, r_i, \alpha_i, \underset{\sim}{y}_\perp\,) \quad . \qquad (4.22)$$

$$\alpha_f = A\,(\, r_i, \alpha_i, \underset{\sim}{y}_\perp\,)$$

V. Kinetic Model for the Two-Particle Distribution

Consider a pair of particle with separation coordinates (r,α). As time progresses (r,α) will change due to interaction with other particles. We shall model this process by including only the three particle interactions described in the previous section. Moreover, we shall model these interactions as discrete, instantaneous events: at randomly chosen times (the pair coordinates (r,α) will be changed (as in (4.22)) to $(r',\alpha') = (R,A)(r,\alpha,\underset{\sim}{y}_{\perp})$ in which $\underset{\sim}{y}_{\perp}$ is chosen independently of (r,α).

The impact coordinates $\underset{\sim}{y}_{\perp}$ may be expressed in polar coordinates as $\underset{\sim}{y}_{\perp} = (\delta,\phi)$. Distant particles have a small effect; so that we restrict δ to $0 < \delta < D$. Since the third particle starts out infinitely far away from the pair, it should be independent and uniformly distributed. Thus we choose $\underset{\sim}{y}_{\perp}$ to be uniformly distributed over the disc $\{|\underset{\sim}{y}_{\perp}| < D\}$ in the plane $\underset{\sim}{y}_{\perp} \cdot \underset{\sim}{v}_0 = 0$.

Let f be a function of (r,α). Its expectation after one collision, starting with (r,α) and ending with (r',α') is

$$\begin{aligned} \langle f(r',\alpha') \rangle_{(r,\alpha)} &= \int_0^{2\pi} \int_0^D f(R,A)(r,\alpha,\delta,\psi)) \; \frac{\delta}{\pi D^2} \; d\delta d\psi \\ &= \int_{\Omega_D} f(r',\alpha') \; \frac{\delta}{\pi D^2} \left| \frac{\partial(R,A)}{\partial(\delta,\psi)} \right|^{-1} dr' d\alpha' \\ &= \int_{\Omega D} f(r',\alpha') \, \tilde{\sigma}(r,\alpha,r',\alpha') \, r'^2 \, dr' d\alpha' \end{aligned}$$

in which (5.1)

$$\tilde{\sigma}(r,\alpha,r',\alpha') = (\pi D^2)^{-1} \, \delta r'^{-2} \, |\partial(R,A)/\partial(\delta,\psi)|^{-1} \tag{5.2}$$

is the collisional transfer density. The range $\Omega_D \subset [0,\infty) \times [0,1]$ for (r',α') depends on D.

The rate $\rho(r,\alpha)$ at which such collisions occur is the rate at which a particle pair (r,α) encounters single particles with $\delta < D$. The relative velocity between pair and single particles is $\underset{\sim}{v}_0(r,\alpha)$ (independent of δ,ψ); the number of particles encountered per distance traveled is $n \pi D^2$. Thus the rate of particle encounters is

$$\rho\,(r,\alpha) = n\pi D^2\, v_0(r,\alpha)\;. \tag{5.3}$$

Multiply $\tilde{\sigma}$ and ρ to get the collisional transfer rate

$$\sigma(r,\alpha,r'\alpha') = n\;\delta\; v_0(r,\alpha) r'^{-2}\; |\partial(R,A)/\partial(\delta,\psi)\;|^{-1}\;, \tag{5.4}$$

which counts the rate at which a pair (r,α) is changed to (r',α') by a collision. Note that the dependence of σ on (r',α') is through (δ,ψ) and the relation $(r',\alpha') = (R,A)(r,\alpha,\delta,\psi)$. Note also that σ is independent of D, although the range Ω_D of (r',α') still depends on D.

Let $p(r,\alpha,t)$ be the probability density for pairs (r,α) as a function of time normalized so that the number of pairs in a set A is $\int_A p(r,\alpha) r^2 dr d\alpha$. The Kolmogorov equation for p is

$$\frac{\partial}{\partial t} p(r,\alpha,t) = \int_{\Omega_D} \{p(r',\alpha',t)\;\sigma\,(r',\alpha',r,\alpha) - p(r,\alpha,t)\sigma(r,\alpha,r',\alpha')\}\; r'^2\; dr' d\alpha' \tag{5.5}$$

Next, we consider scaling of this equation. Since the particles are points, a collision is invariant under scaling, i.e., for $\rho > 0$

$$\begin{aligned} R(\rho r,\alpha,\rho\delta,\psi) &= \rho\; R(r,\alpha,\delta,\psi) \\ A\;(\rho r,\alpha,\rho\delta,\psi) &= A(r,\alpha,\delta,\psi) \end{aligned} \tag{5.6}$$

Thus, if (r,α), $(r'\alpha')$ correscpond to (δ,ψ), then $(\rho r,\alpha)$,$(\rho r',\alpha')$ correspond to $(\rho\delta,\psi)$.

Also

$$v_0(\rho r,\alpha) \;=\; \rho^{-1}\; v_0(r,\alpha). \tag{5.7}$$

It follows that

$$\sigma(\rho r,\alpha,\rho r',\alpha') = \rho^{-3}\sigma(r,\alpha,r'\alpha') \tag{5.8}$$

Using $k = r'/r$, equation (5.5) becomes

$$\frac{\partial}{\partial t} p(r,\alpha,t) = \int_{\Omega_D} \{p(kr,\alpha',t)\ \sigma(1,\alpha',k^{-1},\alpha)\ (rk)^{-3} - p(r,\alpha,t)\sigma(1,\alpha,k,\alpha')\ r^{-3}\} (rk)^2\ d(rk)\ d\alpha'$$

$$= \int_{\Omega_{D/r}} \{p(kr,\alpha',t)\ \sigma(1,\alpha',k^{-1},\alpha)k^{-1} - p(r,\alpha,t)\ \sigma(1,\alpha,k,\alpha')k^2\}\ dk d\alpha' \tag{5.9}$$

Take the limit $D \to \infty$ (assuming the limit exists) for which $\Omega_\infty = [0,\infty) \times [0,1]$. Then the only dependence on r in equation (5.9) is through p. In particular we may look for an equilibrium density by separation of variables as $p(r,\alpha) = r^m q(\alpha)$. The equation for q and m is

$$\int_0^\infty \int_0^1 [k^{m-1} q(\alpha')\ \sigma(1,\alpha',k^{-1},\alpha) - k^3 q(\alpha)\sigma(1,\alpha,k,\alpha')]dk d\alpha' = 0 \tag{5.10}$$

for all α. The exponent m plays the role of a nonlinear eigenvalue. This result must be used with caution: the probability density p is not normalizable because the pairs eventually escape to large r; the integrals in (5.10) are probably separately divergent.

VI. Conclusion

A kinetic theory for collisions between particle pairs and single particles has been derived. It treats the particles as point particles and the three particle interactions as isolated from all other particles. The first of these assumptions requires that $a \ll r$, the second that $r \ll \ell$ (a = particle radius, ℓ = average interparticle spacing, r = pair separation). Therefore, this theory can only be valid if the volume fraction $\phi = (4\pi/3)\ (a/\ell)^3 \ll 1$, and even then is restricted to a range of r values.

The main result is that the equilibrium density for such pairs, although not normalizeable, is separable. It has the form $p(r,\alpha) = r^m q(\alpha)$.

References

1. G.K. Batchelor, "Sedimentation in a dilute suspension of spheres." JFM 52 (1972) 245-268.

2. R.E. Caflisch and J.H.C. Luke, "Variance in the sedimentation speed of a suspension." Physics of Fluids 28 (1985) 759-760.

3. R.E. Caflisch and J. Rubinstein, Lectures on the Mathematical Theory of Multi-Phase Flow (1986) CIMS.

4. J. Happel and H. Brenner, Law Reynolds Number Hydrodynamics (1983) Nijhoff.

5. H. Hasimoto, "On the periodic fundamental solutions of the Stokes equations and their application to viscous flow past a cubic array of spheres." JFM 5 (1959) 317-328.

6. J. Rubinstein, private communication

7. P.G. Saffman, "On the settling speeds of free and fixed suspensions." Studies in Appl. Math. 52 (1973) 115-127

EXPERIMENTS ON SUSPENSIONS OF INTERACTING PARTICLES IN FLUIDS

P. M. CHAIKIN*† AND W. D. DOZIER* AND H. M. LINDSAY*

Abstract.

A system of charged polystyrene spheres in aqueous suspension provides a well characterized way of studying the effects of interparticle interactions on the hydrodynamics of suspensions. We describe the origin of the potential characterizing these interactions and the experiments which are used to test the form of the potential. Techniques, such as light scattering and rheology, have provided a wealth of knowledge about the dynamics of these systems as a function of the density of particles and the strength of the interactions. In many cases the theoretical understanding of the results is lacking and one must treat the results phenomenologically. We summarize some of the more basic questions which remain to be answered.

This conference has dealt with many of the theoretical problems associated with Brownian dynamics of interacting particles. The purpose of the present paper is to bring to your attention some of the experiments that have been performed on systems which fall under the general catagory covered and which may serve to whet your appetites for problems which clearly have solutions (the experimental results) but where the present theory is at best weak if not entirely nonexistent.

Polystyrene spheres ("polyballs") of quite narrow size and surface charge distribution are now readily available from emulsion polymerization techniques.[1] A typical system might consist of 1000 Angstrom particles with ~1000 surface charge groups (sulfonic acid) and a particle density of 10^{13} polyballs/cc. The size polydispersity is commonly 1–2%.

In aqueous suspension the protons dissociate from the sulfonic acid groups and leave the polyballs charged. The system is globally charge neutral with the countercharge associated with the protons solvated in the highly polarizable water. These mobile charges act to screen the electrostatic interactions. In mean field theory the electrostatic potential is then obtained from solutions to the nonlinear Boltzmann-Poisson equation, see Verwey and Overbeek (1948):

$$\nabla^2 \Phi = \rho_0^{-e\Phi/k_B T} \tag{1}$$

where Φ is the electrostatic potential, ρ_0 is the characteristic counterion density, e is the electron charge, k_B is Boltzmann's constant and T the temperature. The analytic form of the solutions to this equation has only been obtained in one dimension. For higher dimensions the conventional approach is to linearize the equations with the result that the Coulomb potential becomes a "screened Coulomb" or Yukawa potential:

$$v(r) = \frac{Z^2 e^2}{\epsilon r} e^{-\kappa r} \qquad \kappa^2 = \frac{4\pi n e^2}{\epsilon k_B T}, \tag{2}$$

where Z is the charge per macro-ion (polyball), ϵ the dielectric constant of the solvant, κ the inverse screening length, and n the total number of singly charged ions in solution. For the case of

*Exxon Research and Engineering Co., Route 22E Annandale, NJ 08801.

†Department of Physics, University of Pennsylvania, Philadelphia, PA 19104.

[1] There is considerable literature on colloidal crystals including the recent reviews by Pieranski (1983) and van Megen and Snook (1984). To get some idea of the various interests in the studies of polyball systems see: Hiltner et al. (1971), Clark et al. (1979), Kose et al. (1973), Schaefer and Ackerson (1975), Ohtsuki et al. (1981), Hachisu et al. (1973) and references therein.

a pure polyball suspension, n is the density of counterions. When additional electrolyte is added, n is the total number of ions (positive plus negative) in solution. (For the more general case ne^2 is replaced by $\Sigma_i n_i (c_i e)^2$ where n_i and c_i are the density and charge of each ionic species.) $v(r)$ is taken as $Ze\Phi(r)$. Linearization of the equations is a quite drastic step, especially in light of the argument of the exponent in equation (1), which is often much larger than unity in the region directly outside the radius of the sphere. However, there has been considerable work which indicates that the Debye-Huckel form of the interaction is appropriate at distances greater than one half of the interparticle separation, but with a renormalized charge $V(r) = Z^* e^2 e^{-\kappa^*}/\epsilon r$, see Alexander et al. (1984). Essentially, the strong field in the nonlinear region near the surface of a polyball attracts a very large countercharge which rapidly reduces the potential until it is of order kT. The additional charge close to the surface effectively renormalizes (reduces) the effective charge at larger distances.

The real test of the appropriateness of the approximations, and their applicability to the experimental system, is the measurement of the forces as a function of the interparticle separation and the inverse screening length, κ^{-1}, controlled by adding electrolyte and hence increasing the number of ions in solution. Fortunately the strength of the interactions between the polyballs is sufficiently strong to cause them to crystallize at room temperature over a rather large range of parameter space. It is therefore possible to measure the interparticle forces by experimentally determining the elastic constants. The first determination of an elastic constant for a polyball or colloidal crystal was the Young's modulus measurement by Crandall and Williams, see Crandall and Williams (1977), an elegant experiment which illustrates some of the clever techniques which can be used to explore these systems. The density of the polystyrene spheres is 1.05 whereas that of water is 1.0. Thus spheres tend to settle and the gravitational pressure is resisted by the modulus of the crystal. The difference in particle density between the top and bottom of a sample is easily probed by Bragg scattering of visible light, since the unit cell dimensions are typically $\sim$0.5 microns, $\sim$ the wavelength of green light. (Colloidal crystals have an appearance much like that of natural opals which are formed by the sedimentation crystalization of 0.5 micron silica particles and also Bragg scatter visible light, see Pieranski (1986).) When the polyballs are suspended in heavy water (D_2O, $\rho = 1.1$) the particle density gradient is reversed. Comparison of the water and heavy water experiments provided a convincing demonstration of the gravitational compression and yielded a measured Young's modulus of $\sim$ 1 dyne/cm^2.

The elastic constants of the colloidal crystals are $\sim$ twelve orders of magnitude less than that of conventional solids. Elastic constants are a measure of the energy density which can be approximated by 1/2 the energy per particle pair times the particle density. The fact that the colloidal crystals are solids at room temperature indicates that the pair energy is $\sim$ two or three orders of magnitude highter than room temperature (as is the case for conventional solids). The great difference in elasticity comes from the particle concentration. In conventional solids we have $\sim$ one particle per Angstrom^3 whereas colloidal crystals have $\sim$ one particle per micron^3. This gives the enormous factor of 10^{12}. For our studies we developed a technique for measuring the shear modulus without using the Bragg scattering (since we also wanted to investigate colloidal glasses, see Lindsay and Chaikin (1983)). The shear modulus was the more relevant parameter for studying the solid phase and, as we shall see, for gaining some information about the liquid state.

The shear modulus of a series of colloidal crystals is shown in figure 1. In figure 1a the electrolyte concentration is being varied and therefore the main effect is on the screening length in the Yukawa potential. In figure 1b the particle concentration is being varied. In practice, we measured the shear modulus for the pure sample from figure 1a, and used this value to calculate the effective

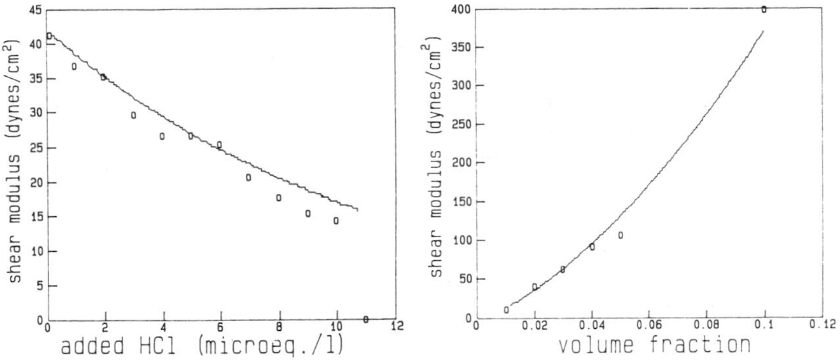

Figure 1. a) Shear modulus as a function of electrolyte concentration for a 4% volume fraction sample of .109 micron polyballs. b) Shear modulus vs. volume fractions for .109 micron polyballs with no added electrolyte. Solid lines are values calculated from equation (2) with a renormalized charge per polyball.

charge per sphere. Additional experiments indicate that the number of surface charge groups are ~5000 whereas the effective charge Z^* is 300. Thus the charge renormalization is substantial. The solid lines indicate the shear modulus that is calculated from the form of the interaction given in equation (1) with the effective charge. The agreement is quite satisfactory and gives us confidence in using equation (1) to describe the interactions. Note however, the abrupt fall in the experimental shear modulus at the electrolyte concentration 11 microequivalents/liter in figure 1a. This corresponds to the first order melting transition. For higher electrolyte concentrations the samples are liquid, as is also evident from the lack of Bragg scattering.

Once the force law is known it is possible to calculate other properties of the solid, such as the crystal structure.[2] For κ small (few screening ions) the interaction is long range and the structure is Body Centered Cubic (BCC) (as previously shown for the classical Wigner crystal resulting from purely Coulombic forces). For a large ionic concentration κ is large, the interaction short range and the close packed Face Centered Cubic (FCC) structure is favored. Only these phases have been observed, in approximately appropriate regimes, but a detailed phase diagram awaits further experimental and theoretical investigation.[3]

Another interesting aspect of these samples is the effect of temperature. The statistical mechanics of the system are obtained from the partition function $\mathfrak{F}$:

$$\mathfrak{F} = \Sigma_{\{r_n\}} e^{-E\{r_n\}/k_B T} = \Sigma\{r_n\} e^{-E'\{r_n\}/\epsilon k_B T} \tag{3}$$

which involves a sum over all possible particle configurations $\{r_n\}$ of the electrostatic interaction energy $E\{r_1, r_2, \ldots, r_n\}$. Even if we go back in our calculations to before we made the Boltzmann-Poisson mean field approximation, so that we include all of the individual ions in $\{r_n\}$, the interactions are screened by the dielectric constant of the solvent. The partition function then depends on $E'\{r_n\}/\epsilon kT \equiv E\{r_n\}/kT$. Water has a high dielectric constant (necessary to solvate the protons and ionize the sulfonic acid groups) because the individual H_2O molecules are dipoles.

[2] See the forthcoming papers by Alexander et al. and Kremer et al.

[3] See the forthcoming paper by Van Winkle and Murray

The temperature dependence of the dielectric constant of free dipoles is a Curie law ($\epsilon \propto 1/T$). Thus ϵkT should be temperature independent and the system would be athermal. Actually the water molecules are not free but rather behave as hindered dipoles. Over the range from freezing to boiling ϵ varies faster than $1/T$, approximately as $T^{-3/2}$. Thus $\epsilon kT \propto T^{-1/2}$ and the temperature dependences are inverted, the samples can melt on cooling (and this has been observed). However, the temperature effects are sufficiently weak over the range 273–373K that temperature is not a good variable. For most experiments we therefore fix the temperature and vary the interaction strength by changing the electrolyte or particle density.

Colloidal suspensions provide an equally unique system for studying the liquid as the solid state. The main advantage over conventional systems is the ability to change the strength and range of the interparticle interactions as well as the particle density. The disadvantage, when trying to understand the physics of simple fluids, is the presence of the suspending fluid, which introduces its own (Brownian) dynamics. Many of the most fundamental properties of the fluid phase are experimentally accessible. The static structure factor, $S(q)$ is observable by light scattering as a function of angle, see Schaefer (1977). The cooperative diffusion constant and the compressibility are usually measured with quasi-elastic light scattering, see Schaefer (1977). The viscosity and its frequency and strain rate dependence are found from conventional rheological measurements, see Krieger (1972), and the self diffusion constant from Forced Rayleigh scattering experiments, see Dozier et al. (1985). $S(q)$ and the cooperative diffusion have been extensively studied by several authors and are semiquantitatively theoretically understood, Schaeffer (1977) and Gruner and Lehman (1982). We will concentrate on the more recent measurements of the viscosity and the self diffusion constant.

The liquid state we are interested in studying is obtained by adding sufficient electrolyte to melt the solid. The interactions are still relatively strong and in particular, with a volume fraction of 1 to several percent, we can be in a regime where the properties (e.g. viscosity and diffusion) of the system are dominated by the electrostatic interactions rather than by the hydrodynamic interactions. We would ideally like to understand some of the properties of simple fluids by studying a fluid of particles suspended in vacuum and interacting by the controllable Yukawa potential. However, the fact that the "polyball fluid" is immersed in another fluid adds a complication. To zeroth order we treat the problem as a "two fluid" system. The viscosity and self diffusion for such a system can be written as

$$\begin{aligned} \eta &= \eta_0 + \eta_1 + 0(\phi) \\ 1/D_s &= 1/D_0 + 1/D_1 + 0(\phi) \end{aligned} \tag{4}$$

where the shear viscosities simply add and the diffusion constants add reciprocally. The two terms in η are η_0, the viscosity of the pure solvent, and η_1, the viscosity of the "polyball fluid" due solely to the interactions between the polyballs. The two terms in D_s are D_0, the Stokes diffusion of a single sphere in the water, and D_1, the diffusion of a single polyball sphere due to interactions with the other spheres in the "polyball fluid." The self diffusion constant is reduced because the Brownian motion is damped by interactions with both fluids. The lowest order corrections to the two fluid model due to hyrodynamic interactions will be linear in the volume fraction of the polyballs with coefficients of order unity (actually 2.5 and -2.7 for viscosity and self diffusion respectively). For the volume fractions used in these studies, $\phi < .08$, the linear corrections should suffice. For most of these studies the observed changes in these transport coefficients are much larger than can be accounted for from hydrodynamic effects and are therefore attributable to the electrostatic interactions.

The viscosity is often described in a phenomenological way, due originally to Maxwell, as an infinite frequency shear modulus G^{∞} multiplied by the time to relax the stress, τ

$$\eta = G^{\infty}\tau \tag{5a.}$$

Since we know the shear modulus in the solid experimentally and theoretically, we can extrapolate the dependence on electrolyte into the liquid state. To try to understand how the Maxwellian relaxation time depends on the stength of the electrostatic interactions we measured the viscosity of the suspension and plotted the results as a function of the calculated/extrapolated shear modulus. The data is shown in figure 2. The parameter being varied is the electrolyte concentration. The finite intercept of the curve is essentially the Einstein correction from hydrodynamic effects (2.5ϕ) plus a small term from electroviscous effects. The slope of the curve is the relaxation time, which surprisingly we found to be a constant, independent of interaction strength.[4]

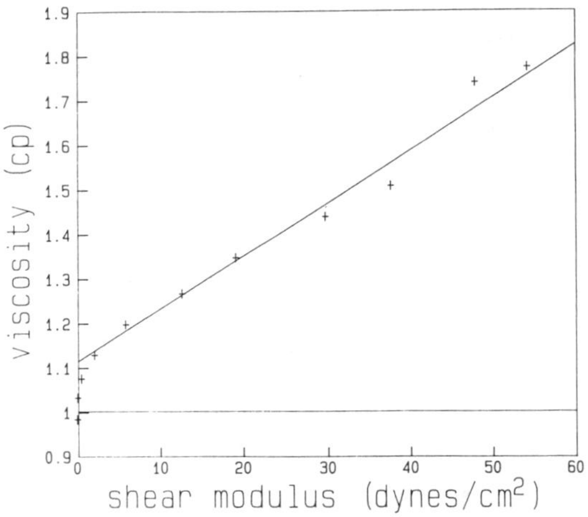

Figure 2. Measured viscosity vs. calculated/extrapolated shear modulus for a 5% volume fraction sample of .109 micron polyballs. The parameter being varied is the electrolyte concentration.

Similar studies on the viscosity of other samples with different size spheres and different volume fractions give the same result, τ independent of the interaction strength. Numerically we found that $\tau = (0.1a)^2/D_0$ for all the samples, where a is the interparticle spacing and D_0 is the free particle Stokes diffusion coefficient. Why does D_0 come in instead of D_s, the self diffusion coefficient for the interacting system, which is strongly dependent on the interactions? And why is the characteristic distance $0.1a$?

When a material is given a sudden strain it responds elastically with a stress field, whether it is solid or liquid. The particles then move to relax the stress. For a solid the stress remains after relaxation, while for a liquid the stress decays after relaxation. The characteristic time for this relaxation can therefore be viewed as the time required for the particles to discover, by their

[4] See the forthcoming paper by Lindsay et al.

motion whether they are in a liquid or a solid. Molecular dynamic simulations, as well as lattice phonon calculations, show that particles in a solid do not displace by more than ∼.1a from their equilibrium lattice positions. These model calculations give some credence to the phenomenological Lindemann melting criteria, see Lindemann (1910), which says that a solid melts whenever the atomic displacements have an RMS value greater than $0.1a$. It is then reasonable to introduce a "Lindemann time", the characteristic time for a particle to traverse the Lindemann distance. At shorter times there is little distinction between the solid and liquid. At longer times the particles may have larger displacements (indicating the liquid state), or the displacements may saturate to a value below $0.1a$ (indicating the solid phase). The Lindemann time is therefore the time in which the system can distinguish between a solid or liquid state, the local melting time. We suggest that this is the appropriate time to describe the stress relaxation.

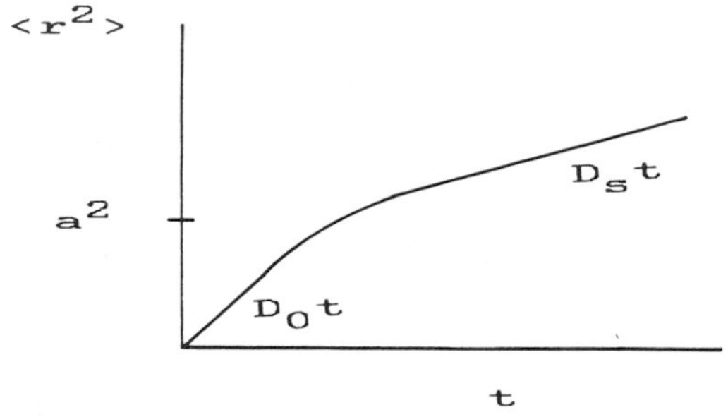

Figure 3. Schematic dependence of the mean square displacement of a Brownian particle in an interacting suspension. The average interparticle separation is a, D_0 is the Stokes diffusion, and D_s is the long time self-diffusion.

In the present system the Lindemann time is set by the Brownian motion, the particles diffuse the distance $0.1a$. We must consider the time dependence of the diffusion constant. The mean square displacement as a function of time is shown schematically in figure 3. At short times the diffusion is dominated by the free diffusion. As the displacement becomes comparable to the interparticle spacing the interparticle interactions play a more important role. The conventional self diffusion comes from the limit where the particle moves many interparticle distances and is slowed down considerably by the difficulty in escaping from the "cage" created by its near neighbors. The crossover between the D_0 and D_s occurs at a distance $\sim a$. Since the Lindemann time requires a displacement of only $0.1a$ it is dominated by the short time diffusion constant D_0.

The above discussion is of course simply hand waving and intuition. The problem has been treated theoretically by Hess and Klein (1983), who use a mode coupling scheme and calculate generalized friction coefficients and a time and wavevector dependent viscosity, $\eta(k,t)$. They find that although the interactions give slightly different time dependences to the viscosity function, the physically observed viscosity involves an integral over all time which essentially washes out even the small differences with interaction strength. Their result gives:

$$\eta_1 = G^{\infty}\frac{a^2}{D_0}\int_0^{\infty} dt'\,\frac{\eta(O,t')}{\eta(O,O)} \tag{5}$$

and numerically the dimensionless integral gives a factor of .01 and therefore the same result as the experiments.

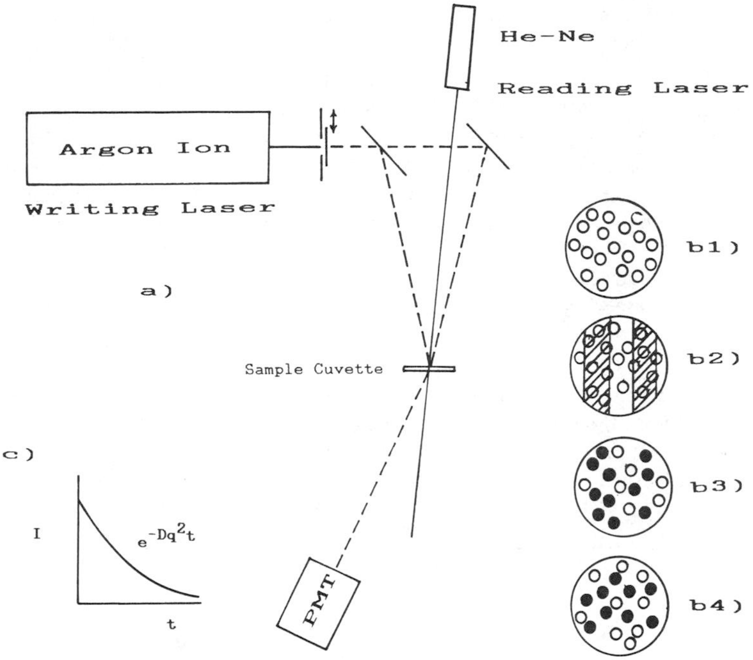

Figure 4. Schematic "Forced Rayleigh Scattering" apparatus. a) Experimental set-up. b1) Polyball sample before exposure to UV radiation. b2) Writing laser shutter is open and interference fringes illuminate sample labelling balls by exciting photochromic dye. b3) With UV off, the instantanious configuration of the dyed polyballs create a diffraction grating which scatters light from the reading laser. b4) Relaxation of the diffraction grating occurs only by the self diffusion of the labelled particles. c) Intensity of light collected by photomultiplier tube, PMT, decays as particles interdiffuse.

There are several interesting points about the relationship $\eta = G\tau$ with $\tau = (0.1a)^2/D_0$. It implies that we can predict the viscosity of the fluid phase of the polyballs from measurements taken entirely in the solid phase. Conventionally, you cannot do that for usual liquids. This works very well for the polyballs due to the simple form for τ given by the Brownian dynamics in the solvent. Whether these simple results can be used for conventional liquids remains to be seen. The mechanism for particle motion over a Lindemann distance depends on the system. For liquid Argon balistic motion at a thermal velocity over a distance (0.1a) gives the relevant τ. Combining this with the shear modulus of the solid phase does, in fact, give the viscosity at $79K$. Of fundamental interest is why such a relation should work at all. The viscosity is related to a stress-stress correlation function whereas, in our use of a Lindemann time, we have used the diffusion constant which is a velocity-velocity correlation function.

We now turn to the question of the self diffusion. The technique which we use to measure the self diffusion is just a slight modification of the usual way self diffusion is defined. We take a set of identical interacting particles and we label one, without changing any of its other properties and measure the distance it diffuses as a function of time. The technique is known as "Forced Rayleigh scattering," see Kreiger (1972) and Hess and Klein (1983). A schematic of our experimental apparatus is shown in figure 4. A sample of polyballs is prepared with a photochromic dye inside the spheres. The dye turns from clear to dark blue when exposed to ultraviolet light. An ultraviolet laser is split into two beams which are recombined on the sample. The difference in laser path lengths produces a sinusoidal interference pattern on the sample. In the high intensity regions the dyed polyballs turn blue, while the less intense region remains white. The ultraviolet beam is left on for a fraction of a second. What remains after the UV is shut off is a series of dark and light stripes formed by the dyed polyballs. This has the form of a diffraction grating. Another laser beam (red) is diffracted off this pattern and the resulting spot reflected onto a photomultiplier tube. The intensity of the diffracted spot is recorded as a function of time. The intensity decays as the diffraction pattern smears out. The only way that this smearing can occur is from the intermixing of the dark and white spheres by self diffusion. The exponential decay time for the diffracted beam is $D_s q^2$ where q is the wavevector of the imposed grating.

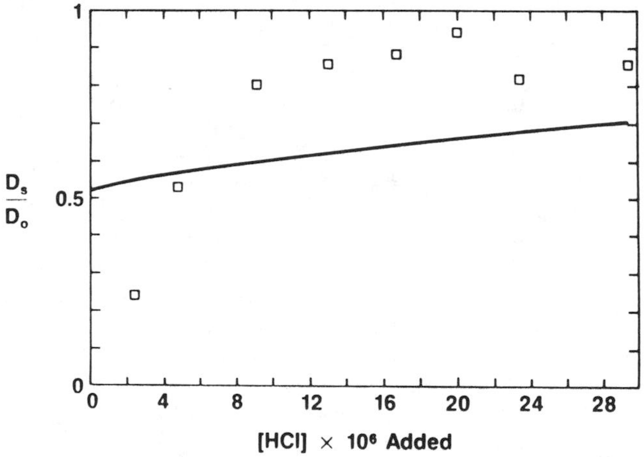

Figure 5. Self diffusion for a sample of .109 micron polyballs, 1% volume fraction vs. electrolyte concentration. Solid line is from theory of Hess and Klein (1983).

The self diffusion coefficient for a 1% sample of .109 μ polyballs is shown in figure 5 as a function of the electrolyte concentration. For very large electrolyte concentration, hence weakly interacting particles, the diffusion approaches the Stokes value. As the electrolyte concentration is reduced the particles interact more strongly and the self diffusion is reduced until at 2 μmoles/liter it drops by at least three orders of magnitude indicating the freezing transition to the solid. By adding small amounts of electrolyte or pure suspension it is possible to move very carefully through this transition and confirm that it is abrupt and first order. We also see that it is possible to have a significant variation in the diffusion constant ($\sim$ a factor of 10), see Dozier et al. (1985).

Theoretically corrections to the Stokes diffusion for non-interacting particles have been pertubatively treated for weak interactions. Among the more recent treatments is that of Hess and Klein (1983). Their theory is shown as the solid line on figure 5. While the highly screened limit is fairly well treated, substantial reductions in the diffusion coefficient are beyond the scope of this and most present theories. We are in the limit where the electrostatic interactions dominate over the Brownian motion in the solvent. It is within this context that we treat the "two fluid model", equation (4), with the main effects coming from the polyball liquid. Unfortunately there also is not a good theory for the diffusion coefficient for a simple liquid. Thus the best that we can do at present is to see whether the behavior of the polyball fluid is consistent with the relations found in simple liquids.

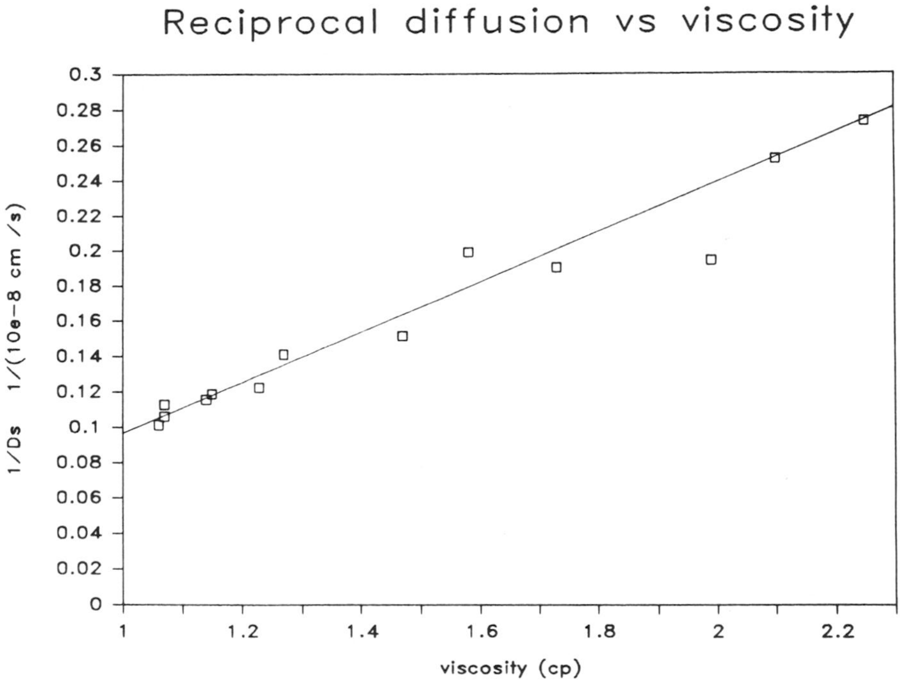

Figure 6. Inverse of the self-diffusion vs. viscosity for a 5% sample of 0.038 micron polyballs. Electrolyte concentration is being varied.

Phenomenologically it has long been known that the diffusion constant for an atom or molecule in a simple liquid is well described by a Stokes like equation

$$D_s = \frac{k_B T}{A \eta R} \tag{6}$$

carried to the atomic limit where the hydrodynamic derivation should no longer be applicable. In equation (6), η is the viscosity of the bulk fluid, R is the particle diameter, and A is a constant which varies from 2π to 4π but is most often 3.3π. To see whether this applies to the polyball fluid, we have used the two fluid equations to determine the viscosity and diffusion constants of this fluid (subtracting out the terms due to the water). We then plot the reciprocal of the viscosity versus the self diffusion constant in figure 6. The straight line through the data shows that the proportionality holds, which is interesting in itself. However, the proportionality constant works out to be 3.3π when we take 1/2 the interparticle distance instead of the particle diameter in equation (6). In the case of a simple atomic or molecular liquid, there is no difference between the particle diameter and the interparticle spacing. This is because for the simple fluid the particle diameter is not taken as the nuclear radius but rather the effective hard sphere radius which characterizes the interparticle interaction and sets the spacing. Thus for the polyball fluid it is inappropriate to use the polyball diameter. The equivalent to the atomic diameter is the interparticle spacing (we are dealing with rather long range purely repulsive interactions). In order to test this idea we measured the viscosity and the diffusion constant for a series of samples where the volume fraction changed by a factor of 8. Using a fixed value for R rather than the interparticle spacing would introduce a factor of two error in the predicted value of the diffusion constant. The data shown in figure 7 support our idea that the appropriate R is 1/2 the interparticle spacing.

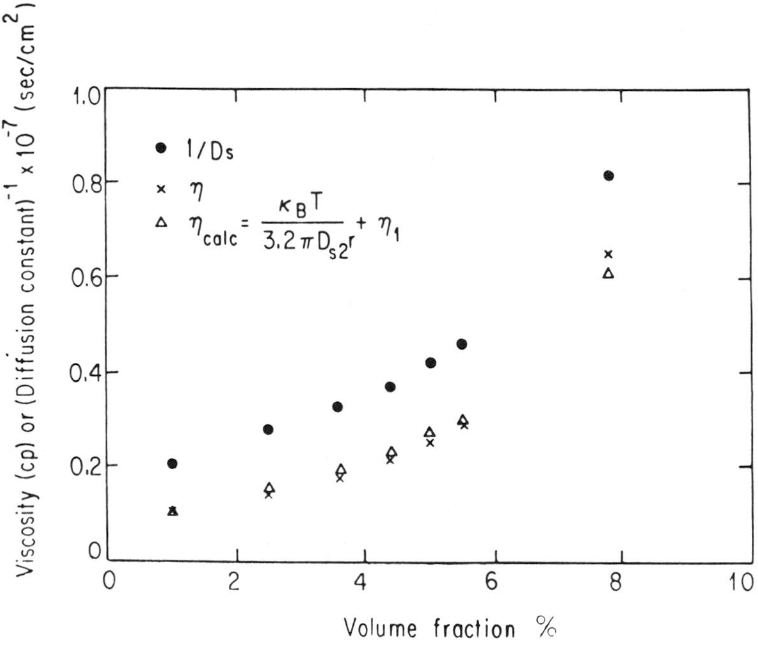

Figure 7. Self diffusion constant, ., and viscosity, x, for a series of 0.038 micron polyball samples vs. volume fraction. Viscosity calculated from the diffusion constant, Δ, is also shown.

Of course the major theoretical problem in the above discussion is why the phenomenological relation between the diffusion constant and the viscosity works at all. It has been known to hold for many decades but nobody knows why. It is again a problem in relating a velocity-velocity correlation function to a stress-stress correlation function.

There are additional experiments on the polyball system which require theoretical understanding on a very basic level. One of the most dramatic is the phenomena of "shear-melting" first discovered by Hoffman (1974) and Ackerson and Clark (1981,1984). When a crystalline sample is subjected to a steady-state shear, it will flow (due probably to disclocation motion at first) but remain in the crystalline state. As is the case for most solids the flow is highly non-Newtonian and shear thinning. However, as the shear rate is further increased there comes a point at which the stress jumps and the crystalline phase disappears as evidenced by the lack of Bragg scattering, see Pohl et al. (1973) and Rondelez et al. (1978). Presently this transition is treated variously as resulting from a mechanical or hydrodynamic instability, see Ackerson and Clark (1981) and Hoffman (1972), or as the result of a small nonequilibrium deviation from a statistical mechanical transition, see Lindsay and Chaikin (1985).

In conclusion the charged colloids provide a unique system for the experimental investigation of a wide variety of phenomena associated with solids and liquids. The main advantages are the basic understanding and controllability of the interactions and the particle density. They should be ideal systems for testing theoretical results. We have presented some of our results on the behavior of the liquid phase and suggested that it may be regarded as a simple liquid. We have shown that the viscosity can be obtained by the usual Maxwell relation $\eta = G\tau$ with a new interpretation of the stress relaxation time in terms of a Lindemann time. The theoretical problems for which we lack a basic understanding are the relationship between the transport coefficients which depend on velocity-velocity correlation function and stress-stress correlation functions, and a way of calculating these quantities in strongly interacting systems.

REFERENCES

[1] B. J. Ackerson and N. A. Clark, Phys. Rev. Lett., 46 (1981), p. 123.
[2] B. J. Ackerson and N. A. Clark, Phys. Rev., A30 (1984), p. 906.
[3] S. Alexander, P. M. Chaikin, P. Grant, G. J. Morales, P. Pincus and D. Hone, J. Chem. Phys., 80 (1984), p. 5776.
[4] S. Alexander, P. M. Chaikain, D. Hone, P. Pincus and D. Schaefer, (to appear).
[5] N. A. Clark, B. J. Ackerson, and T. W. Taylor, J. Phys. (Paris) Colloq., C3, 46 (1985), p. 137.
[6] N. A. Clark, A. J. Hurd, and B. J. Ackerson, Nature, 281 (1979), p. 5726.
[7] R. S. Crandall and R. Williams, Science, 198 (1977), p. 293.
[8] W. D. Dozier, H. M. Lindsay, and P. M. Chaikin, J. Physique, Colloque, 46, C3 (1985).
[9] F. Gruner and w. Lehmann, J. Phys., A12 (1979), p. L303.
[10] F. Gruner and W. Lehmann, J. Phys., A15 (1982), p. 2847.
[11] S. Hachisu, Y. Kobayashi and A. Kose, J. Colloid Interface Sci., 42 (1973), p. 342.
[12] W. Hess and R. Klein, Adv. Phys., 32 (1983), p. 173.
[13] P. A. Hiltner, Y. S. Papir and I. M. Kreiger, J. Phys. Chem., 75 (1971), p. 1881.
[14] R. L. Hoffman, Trans. Soc. Rheology, 16 (1972), p. 155.
[15] R. L. Hoffman, J. Coll. Int. Sci., 46 (1974), p. 491.
[16] R. Klein and W. Hess, Farad. Disc. Chem. Soc., 76 (1983).
[17] A. Kose, T. Osake, Y. Kobayashi, K. Takano, and S. Hachisu, J. Coll. Int. Sci., 44 (1973), p. 330.
[18] K. Kremer, M. Robbins, and G. Grest, (to appear).
[19] I. M. Krieger, Adv. Colloid and Interface Science, 3 (1972), p. 111.
[20] F. A. Lindemann, Z. Phys., 11 (1910), p. 609.
[21] H. M. Lindsay and P. M. Chaikin, J. Chem. Phys., 76 (1983), p. 3774.
[22] H. M. Lindsay and P. M. Chaikin, J. Phys. (Paris) Colloq., C3, 46 (1985), p. 269.
[23] H. M. Lindsay, W. D. Dozier, P. M. Chaikin, R. Klein, and W. Hess, J. Phys. C (to appear).
[24] W. van Megen and I. Snook, Adv. Coll. Int. Sci., 21 (1984), p. 119.
[25] T. Ohtsuki, A. Kishimoto, S. Mitaku, and K. Okeno, Jpn. J. Appl. Phys., 20 (1981), p. 509.
[26] P. Pieranski, Contemp. Phys., 24 (1983), p. 25.
[27] P. Pieranski, La Recherche, 17 (1986), p. 312.
[28] D. W. Pohl, S. E. Schwarz and V. Irniger, Phys. Rev. Lett., 31 (1973), p. 32.

[29] F. Rondelez, H. Hervet, and W. Urbach, Chem. Phys. Lett., 41 (1978), p. 138.
[30] S. Ramaswamy and S. R. Renn, Phys. Rev. Lett., 56 (1986), p. 945.
[31] D. W. Schaeffer, J. Chem. Phys., 66 (1977), p. 3980.
[32] D. W. Schaefer and B. J. Ackerson, Phys. Rev. Lett., 35 (1975), p. 1448.
[33] D. H. Van Winkle and C. A. Murray, (to appear).
[34] E. J. W. Verwey and J. Th. G. Overbeek, *Theory of Stability of Lyophobic Colloids*, Elsevier, Amsterdam, 1948.

STOCHASTIC MODELS OF PARALLEL SYSTEMS FOR GLOBAL OPTIMIZATION

D.A. Dawson
Department of Mathematics & Statistics
Carleton University
Ottawa, Canada, K1S 5B6

ABSTRACT

Random search methods provide a natural approach to the problem of finding a global extremum of a function which has many critical points. Recently some basic ideas of statistical physics have been exploited in the development of the "annealing algorithm" which has been successfully applied to certain combinatorial optimization problems. The objective of this paper is to consider an interacting particle system model of a many searcher system and to reformulate some simple ideas from statistical physics in this context.

1. INTRODUCTION TO THE PROBLEM

Consider the problem of finding a global maximum of a smooth function $V(\cdot)$ defined on $S \subseteq R^d$. If V has a unique global maximum at x_* and ∇V satisfies certain natural conditions, then the solution of the differential equation

$$dx(t)/dt = \nabla V(x(t)) \tag{1.1}$$

yields

$$x_* = \lim_{t \to \infty} x(t)$$

for any initial value $x(0)$.

When the function V has multiple local maxima or flat regions (mesas), then this method is no longer appropriate. In such cases random search methods provide an alternative. For example if the domain S is a bounded region then a standard method consists of generating finitely many random variables which are uniformly distributed on S and then determining the maximum in this random sample (cf. Dorea (1986) for convergence results for this method). For many optimization problems it is natural to combine the ideas of local improvement illustrated by (1.1) together with random search. Such a method which is known as the "annealing algorithm" was proposed in the context of combinatorial optimization by Černý (1982) and Kirkpatrick et al. (1983). This discrete time method is based on the "Metropolis algorithm" which leads to a Markov chain with the equilibrium distribution

$$\mu(dx) = Z^{-1} e^{V(x)/kT} dx \tag{1.2}$$

where k is a constant, T is the "temperature" and Z is the normalizing constant. If T is then slowly lowered to zero (annealing), then it is reasonable to expect that the system approaches the global maximum of V as time passes. However it is necessary to cool the system sufficiently slowly so that it can escape from local maxima which are near the global maximum.

To illustrate these ideas we briefly review the recent work of Geman and Hwang (1985). Let $S = [0,1]^d$ and consider the stochastic differential equation

$$dx(t) = \nabla V(x(t))dt + \sqrt{2}\sigma(t)dw(t) \tag{1.3}$$

where $w(\cdot)$ is a standard Brownian motion with reflection on the boundary of S. For $\sigma(\cdot) = \sigma$, then the equilibrium distribution is

$$\mu_\sigma(dx) = Z^{-1} e^{V(x)/\sigma^2} dx . \tag{1.4}$$

It is natural to assume that $\mu_\sigma \longrightarrow \mu_0$ as $\sigma \longrightarrow 0$ and that μ_0 is supported by the set

$$M = \{x : V(x) = \max_y V(y)\} .$$

Under certain regularity conditions on V, Geman and Hwang (1985) have identified the appropriate "annealing schedule" $\sigma(\cdot)$ for this problem. In particular they proved that if $\sigma^2(t) \geq c/\log(1+t)$, and c is sufficiently large, then

$$P(x(t) \in \cdot \,|\, x(0) = x) \longrightarrow \mu_0(\cdot)$$

for all $x \in S$. The idea is that this system can escape from any finite local (or global) maximum for arbitrarily large times. In fact it will jump between local maxima but will spend an increasing proportion of time near the global maximum. Nevertheless this system can be temporarily trapped near a non-global local maximum for long times. In fact the theory of Freidlin and Wentzell (1984) implies that for a constant small temperature σ^2 the mean exit time is of the order of $\exp(\Delta V/\sigma^2)$ where ΔV denotes the "height" of the local maximum.

For many search problems (e.g. search for ship-wreck survivors) large regions of S contain no information (mesas) and in addition it is important to minimize search time. In such cases the obvious method is to employ many searchers. The simplest multiple searcher system is one in which there are N independent searchers each carrying out a search according to a random search of the type described above. If some type of communication exists between the searchers, then the system becomes an "interacting stochastic system". This raises the possibility that ideas from statistical physics can also be useful in the study of such parallel random search methods.

2. FORMULATION OF SOME PARALLEL SEARCH PROBLEMS AND MODELS

2.1 INTRODUCTION

Consider a system of N searchers. For $j = 1,\dots,N$, $x_j(t) \in S$ denotes the location of the jth searcher at time t. An *estimate* of the target location (position of a global maximum of V) at time t could be obtained by one of the following:

(a) sampling a searcher at random, or

(b) taking the average position of the searchers

$$\bar{x}(t) = N^{-1} \sum_{j=1}^{N} x_j(t).$$

The time until the search is within ε of its objective is defined by

$$\tau_S^{\varepsilon} = \inf\{t : N^{-1} \sum_{j=1}^{N} V(x_j(t)) \geq V_{max} - \varepsilon\}$$

$$\tau_A^{\varepsilon} = \inf\{t : \bar{x}(t) \in M^{\varepsilon}\}$$

where

$$V_{max} = \max_x V(x), \quad \text{and} \quad M^{\varepsilon} = \{x : V(x) \geq V_{max} - \varepsilon\}.$$

The overall objective of the parallel search problem is to find an algorithm which yields $E(\tau_S)$ or $E(\tau_A)$ as small as possible for a randomly chosen initial condition. Such an algorithm must take into account the constraints on the motion of the searchers (e.g. maximum speed) and also the nature of the information available to a given searcher.

The two heuristic ideas which motivate the development of parallel search algorithms are:

(i) the use of many random searchers to cover more territory in a fixed interval of time and thus to decrease the time required to find promising search regions,

(ii) the adaptive reallocation of resources to put more search resources in the promising search regions.

In order for heuristic (ii) to be relevant it is necessary that the function V have some type of "large scale regularity". However the objective is to produce methods which share with the annealing algorithm the ability to escape from local maxima of the objective function V.

2.2 THE SIMPLE WEIGHTED MEAN FIELD DRIFT INTERACTION

Let $S = R^1$. A simple model of cooperative search is one in which each searcher (particle) exerts an attractive force on the other particles and that the strength of the force exerted by the jth particle at time j depends on the value $V(x_j(t))$. In other words particles have a tendency to move towards other particles whose current locations have the largest V-values. We model this with the following system of stochastic differential equations: for $j = 1,\dots,N$,

$$dx_j(t) = \nabla V(x_j(t))dt + \sigma dw_j(t) + \rho(\bar{\bar{x}}(t) - |\bar{V}(t)|x_j(t))dt \tag{2.1}$$

where $\rho,\sigma > 0$, $V \leq 0$ (this can be assumed without loss of generality),

$$\bar{\bar{x}}(t) = N^{-1}\sum_{j=1}^{N} x_j(t)V(x_j(t)), \quad \text{and}$$

$$\bar{V}(t) = N^{-1}\sum_{j=1}^{N} V(x_j(t)).$$

The analysis of this system is based on reformulating (2.1) in terms of the associated empirical measure process

$$X_N(t) = N^{-1}\sum_{j=1}^{N} \delta_{x_j(t)} .$$

The process X_N is a $M_1(R^1)$-valued Markov diffusion process.

In order to understand the behavior of this and other interacting particle models for large N it is useful to study the *mean-field limit* $(N \to \infty)$ (cf. Gärtner (1985), Léonard (1984), McKean (1969), Oëlschläger (1984)). In this case the motion of a *tagged searcher* is given by the nonlinear stochastic differential equation

$$dx(t) = \nabla V(x(t))dt + \sqrt{2}\,\sigma dw(t) + \rho[E(x(t)V(x(t)))-x(t)|E(V(x(t))|]dt \tag{2.3}$$

where $E(f(x(t))$ denotes the expectation of $f(x(t))$ under the probability law of the solution of (2.3). This means that the limiting distribution of searchers is given by a weak solution of the nonlinear McKean-Vlasov equation:

$$\partial p(t,x)/\partial t = \Delta p(t,x) - \partial/\partial x(\nabla V(x)p(t,x))$$
$$-\rho\partial/\partial x\left([\int yV(y)p(t,y)dy - x|\int V(y)p(t,y)dy|]p(t,x)\right) \tag{2.4}$$

where $\int_A p(t,x)dx$ denotes the limiting mass of searchers in A at time t .

Example 2.2.1. Single Maximum

Let $V(x) = -x^2$. Then (2.3) becomes

$$dx(t) = -x(t)dt + \sqrt{2}\,\sigma dw(t) + \rho(-m_3(t) - m_2(t)x(t))dt \tag{2.5}$$

where $m_n(t) = E(x^n(t))$.

The invariant probability densities for the resulting form of Equation (2.4) are of the form

$$p(m_2,m_3,x) = Z^{-1}\left(\sigma^{-2}[-x^2(1+\rho m_2)-\rho m_3 x]\right) . \tag{2.6}$$

The possible values of m_2, m_3 are obtained by solving the self-consistency equations

$$\int x^2 p(m_2,m_3,x)\,dx = m_2 ,$$

$$\int x^3 p(m_2,m_3,x)\,dx = m_3 .$$

For example one solution is given by

$$m_3 = 0,\ m_2 = \frac{-1 \pm \sqrt{1+2\sigma^2\rho}}{2\rho} .$$

Note that $m_2 \to 0$ as $\rho \to \infty$; this suggests that increasing the interaction strength ρ plays a role analogous to that of decreasing the temperature in the annealing algorithm.

Example 2.2.2. Two Local Maxima.

Consider the one parameter family

$$V(x) = cx + x^2 - x^4 .$$

For $|c| < c_0$ $(= 4/3\sqrt{6})$, this function has two local maxima. The corresponding invariant probability densities have the form

$$p(m_1,m_2,m_3,m_4,m_5;x) = Z^{-1}\exp\left(\sigma^{-2}[-x^4+cx+x^2+g(x,m_1,m_2,m_3,m_4,m_5)]\right)$$

where $m_1,\ldots,m_5$ are obtained by solving a system of five self-consistency equations. As the parameters σ, c, ρ are varied this system can exhibit bifurcation of multiple equilibria. In particular there can exist equilibrium densities concentrated near a (non-global) local maximum. Since the empirical distribution of searchers follows the McKean-Vlasov equation for long periods of time when N is large, this means that in such a case the finite searcher system can get trapped at a local maximum.

3. TRAPPING AT A LOCAL MAXIMUM

In this section we analyse in detail the phenomenon of trapping in a local maximum for a simplified model in which we replace the weighted mean-field interaction with a simple mean-field interaction. However the methods involved are not restricted to this special case and analogous results are to be expected in the weighted mean-field case. Since these results appear in detail elsewhere we summarize the main ideas.

Let $V(x) = x^2-x^4+cx$ and consider the system of stochastic differential equations: for $j = 1,\dots,N$,

$$dx_j(t) = \nabla V(x_j(t))dt + \sqrt{2}\,\sigma dw_j(t) + \rho(\bar{x}(t)-x_j(t))dt \tag{3.1}$$

where $\bar{x}(t) = N^{-1}\sum_{j=1}^{N} x_j(t)$.

Let $M^* = \{\mu \in M_1(R^1) : \int x^4\mu(dx) < \infty\}$. The empirical measure process defined by (2.2) is a M^*-valued diffusion process with generator:

$$G_N F(\mu) = G^{(1)}F(\mu) + N^{-1}G^{(2)}F(\mu) , \tag{3.2}$$

where $G^{(1)}F(\mu) = \int L^{\mu}\left(\frac{\delta F(\mu)}{\delta\mu(x)}\right)\mu(dx)$,

$$L^{\mu}f(x) = \sigma^2\Delta f + (\partial V/\partial x)\partial f/\partial x - \rho\left\{[x - \int y\mu(dy)]\partial f/\partial x\right\} , \quad \text{and}$$

$$G^{(2)}F(\mu) = \sigma^2\iint \frac{\partial^2}{\partial x\partial y}\left(\frac{\delta^2 F(\mu)}{\delta\mu(x)\delta\mu(y)}\right)\delta_y(dx)\mu(dy) .$$

In the mean-field limit, $N \to \infty$, (cf. Dawson (1983)),

$$X_N(.) \to X_\infty(.) \quad \{\text{weak convergence of processes}\}$$

where $X_\infty(t,dx) = \mu(t)$ where $\mu(.)$ is given by the (deterministic) weak solution of the equation:

$$\frac{d\mu}{dt} = L^{\mu(t)^*}\mu(t) \tag{3.3}$$

where L^{μ^*} denotes the adjoint of L^{μ}.

The invariant densities are given by

$$P_a(x) = Z^{-1}\exp(\sigma^{-2}[-x^4 + (1-\rho)x^2 + ax]) \tag{3.4}$$

where $a = c + \rho\int xp_a(x)dx$.

If we define $m(a) = \int xp_a(x)dx$, then the self-consistency equation becomes:

$$m_1 = m(c + \rho m_1). \tag{3.5}$$

Since the function $m(a)$ is concave for $a \geq 0$ and an odd function, we conclude that

(i) if $c = 0$, $m_1 = 0$ is always a solution, an additional symmetric pair $m_\pm$ exists if and only if $\rho > \rho_c = (m'(0))^{-1}$, (yielding invariant densities $\mu_\pm$),

(ii) when $|c|$ is sufficiently large there is a unique solution, μ_g,

(iii) when $\rho > \rho_c$ and $c \neq 0$ but $|c|$ is sufficiently small there exists a pair of stable solutions μ_g, μ_ℓ and one unstable solution μ_u.

The solution μ_g denotes the invariant measure concentrated near the global maximum and μ_ℓ denotes the invariant measure concentrated near the (non-global) local maximum. In other words if we increase the strength of the interaction (or decrease the temperature σ^2), then the mean-field limit gets trapped at a local maximum. This means that the attractive force between the large number of particles at the local maximum overrides the information sent by a small number of particles at the global maximum.

In the rest of this section we consider the long-time behavior of the (finite) N-searcher empirical process X_N. The process $X_N(.)$ is ergodic and has an invariant probability measure Π_N on M^*. It follows from large deviation theory (cf. Gärtner (1985)) that $\{\Pi_N : N \geq 1\}$ forms a large deviation family with *action* (or *entropy*) functional F. This means that for fixed $T > 0$ and $A \subset C([0,T],M^*)$,

$$-\inf\{F(\mu) : \mu \in A^{o}\} \leq \liminf_{n\to\infty} N^{-1} \log \Pi_N(A) \tag{3.6}$$

$$\leq \limsup_{n\to\infty} N^{-1} \log \Pi_N(A)$$

$$\leq -\inf\{F(\mu) : \mu \in \overline{A}\}$$

where A^{o}, $\overline{A}$ denote the interior and closure of A, respectively. In this application F is given by

$$F(\nu) = \int \log\left(\frac{d\nu}{d\mu_{\sigma,\rho}}\right)d\nu - (\rho/2\sigma^2)m(\nu)^2 \tag{3.7}$$

where $m(\nu) = \int x\nu(dx)$, and

$$\mu_{\sigma,\rho}(dx) = Z^{-1} \exp(\sigma^{-2}[V(x)-\tfrac{1}{2}\rho x^2])dx.$$

Formally the mean-field limit $\mu(.)$ can also be characterized as the solution of the equation:

$$\frac{d\mu(t)}{dt} = -\nabla F(\mu(t)).$$

The action functional F is useful in describing the relation between the stability properties of the mean-field limit $\mu(.)$ and the long-time behavior of the N-searcher empirical process $X_N(.)$. In particular we consider the problem of the exit from the domain of attraction of the stable equilibria (cf. Shiino (1985) for a study of the global stability properties of the equilibria). In order to carry out this analysis it is necessary to develop the analogue of the Freidlin-Wentzell theory for the above system when we consider the process X_N with generator given by (3.2) as a small random perturbation of the mean-field limit. The main results are summarized in the following theorem.

THEOREM 3.1. (Dawson and Gärtner (1986)).

(a) Assume that $X_N(0) \to \mu_0 \in M^*$ and that $P^N_{\mu_N}$ is the probability measure on $C([0,T],M^*)$ associated with the process X_N. Then $\{P^N_{\mu_N} : N \geq 1\}$ forms a large deviation family with action functional $S^T(.)$. The action functional is given by:

$$S^T(\mu(.)) = \int_0^T \|\mu'(t) - L^{\mu(t)*}\mu(t)\|_{\mu(t)} dt$$

$$= +\infty \quad \text{if the above expression is not well-defined,}$$

where $\mu'(t) = \dfrac{d\mu(t)}{dt}$, and

$$\|\nu\|_\mu = \sup_{\phi \in \mathcal{D}} \left(\frac{(\int \phi(x)\nu(dx))^2}{\sigma^2 \int |\nabla\phi|^2 \mu(dx)} \right), \quad |\nabla\phi|^2 = [\partial\phi/\partial x]^2.$$

(b) For μ in the domain of attraction of μ_α, $\alpha = \ell$ or g, the quasi-potential

$$Q_\alpha(\nu) := \inf\{S^T(\mu) : \mu(0) = \mu_\alpha, \ \mu(T) = \nu, \ T \geq 0\}$$

$$= F(\nu) - F(\mu_\alpha), \quad \alpha = \ell \text{ or } g.$$

As an application of these results we obtain the following analogue of the Freidlin-Wentzell exit result. Let U_u be a neighbourhood of μ_u and

$$\tau_u = \inf\{t : X_N(t) \in U_u\}.$$

Assume that $X_N(0) \to \mu_\alpha$, $\alpha = \ell$ or g. Then

$$\lim_{N\to\infty} N^{-1} \log E_N(\tau_u \mid X_N(0)) = Q_\alpha,$$

where $Q_\alpha = \inf_{\nu \in U_u} [F(\nu) - F(\mu_\alpha)] \geq 0$

and E_N denotes expectation with respect to the N-searcher system.

In other words, if the mean-field limit has an invariant measure μ_ℓ concentrated near the local maximum, then the N-searcher system gets trapped there for a time of the order of (e^{NQ_α}). This search system has the undesirable characteristic that as the number of searchers is increased it gets trapped for longer periods near a local maximum. This means that the simple mean-field interaction is too strong and that other interaction structures must be considered.

4. A HIERARCHICAL APPROACH

We begin by reviewing the performance of the mean-field model with respect to the criteria involving $E_N(\tau_S^\varepsilon)$ and $E_N(\tau_A^\varepsilon)$. For the test class of V's considered in Section 3 and $\varepsilon > 0$ there exists ρ such that

$$\lim_{N\to\infty} P_{\mu_N}^N(\tau_A^\varepsilon < \infty) = 1.$$

However $\bar{x}(\cdot)$ can get stuck close to any non-global local maximum for time periods of order e^{NQ_ℓ} and therefore $E_N(\tau_S^\varepsilon)$ and $E_N(\tau_A^\varepsilon)$ can be very large.

The purpose of this section is to propose a hierarchically structured interaction such that given $\varepsilon > 0$ we can design a system for which $E_N(\tau_S^\varepsilon) \leq K < \infty$ for all N and initial measures. (Note that this property is not achieved by the annealing algorithm.)

Consider a finite two level system which consists of M search groups each consisting of N searchers. For $i = 1,\ldots,M$ and $j = 1,\ldots,N$ let $x_{ij}(t)$ denote the location of the jth searcher in the ith group. Let

$$X_i^N(t) = N^{-1} \sum_{j=1}^{N} \delta_{x_{ij}(t)} \in M_1(R^1), \quad \text{and}$$

$$X^{M,N}(t) = M^{-1} \sum_{i=1}^{M} \delta_{X_i^N(t)} \in M_2 = M_1(M_1(R^1)).$$

The multilevel system to be considered is given by

$$\begin{aligned} dx_{ij}(t) = {} & [\nabla V(x_{ij}(t)) + \rho(N^{-1}\sum_{k=1}^{N} x_{ij}(t) - x_{ij}(t)) \\ & + \theta N^{-\alpha}(MN)^{-1}\sum_{i=1}^{M}\sum_{j=1}^{N} x_{ij}(t) - N^{-1}\sum_{j=1}^{N} x_{ij}(t))]dt \\ & + \sqrt{2}\,\sigma dw_{ij}(t) \end{aligned} \tag{4.1}$$

where $\rho,\theta,\sigma > 0$ and $\alpha \geq 0$. The term with coefficient $\theta N^{-\alpha}$ is the level two interaction.

The generator of the M_2-valued process $X^{M,N}$ has the form:

$$G_{M,N} = G_0 + N^{-1}G_1 + M^{-1}G_2 + (MN)^{-1}G_3 \,. \tag{4.2}$$

where $G_0F(\nu) = \int L^\mu\left(\frac{\delta F(\mu)}{\delta\nu(\mu)}\right)\nu(d\mu)$, with L^μ defined as in Section 3,

$$\frac{\delta F(\nu)}{\delta\nu(\mu)} = \frac{d}{d\varepsilon} F(\nu+\varepsilon\delta_\mu)\Big|_{\varepsilon=0}, \text{ and}$$

$$G_1F(\nu) = \sigma^2 \iint \left(\partial/\partial x\left[\frac{\delta}{\delta\mu(x)}\left(\frac{\delta F(\nu)}{\delta\nu(\mu)}\right)\right]\right)^2 \mu(dx)\nu(d\mu)$$

and G_2, G_3 are described below. Although we are interested in the behavior of this system for finite M and N, it is of interest to study the behavior as M and/or $N\to\infty$.

PROPOSITION 4.1 (The $M\to\infty$ Law of Large Numbers).

As $M\to\infty$ the processes $X^{M,N}(.)$ converge in the sense of M_2-valued processes to a process of the form $Y^N(.) = \delta_{X_N(.)}$ where the limit process $\{X_N(.)\}$ is characterized as the unique solution to the $M_1(R^1)$-valued martingale problem associated with the nonlinear operator G_N^ν defined as follows. The class of test functions are of the form $F_{f,g}(\mu) = f(\langle\mu,g\rangle)$ where $\langle\mu,g\rangle = \left(\int g(x)\mu(dx)\right)$ with $f\in C_b^2(R^1)$, $g\in C_K^2(R^1)$. Then the limiting nonlinear generator is given by

$$G_N^\nu F_{f,g}(\mu) = f'(\langle\mu,g\rangle)[\langle\mu,L^\nu(\mu)g\rangle] + \sigma^2N^{-1}f''(\langle\mu,g\rangle)\langle\mu,|\nabla g|^2\rangle \tag{4.3}$$

where

$$L^\nu(\mu)g = \sigma^2\Delta g + (\partial V/\partial x)\partial g/\partial x - \rho[(x-\int y\mu(dy))\partial g/\partial x]$$
$$-\theta N^{-\alpha}[\{\int y\mu(dy) - \iint y\eta(dy)\nu(d\eta)\}\partial g/\partial x].$$

REMARK. This result is actually a special case of the general law of large numbers for interacting diffusions (cf. Gärtner (1985), Léonard (1984), Oëlschläger (1984)) applied to the family of M symmetrically interacting R^N-valued diffusions defined by (4.1).

Proposition 4.1 can be reinterpreted as yielding the following nonlinear stochastic differential equation for a *tagged search group:* for $j = 1,\ldots,N$,

$$dx_j(t) = \nabla V(x_j(t))dt + \sqrt{2}\,\sigma dw_j(t) + \rho(\bar{x}(t)-x_j(t))dt \tag{4.4}$$
$$+\theta N^{-\alpha}[E(\bar{x}(t)) - \bar{x}(t)]dt$$

where $\bar{x}(t) = N^{-1}\sum_{j=1}^{N} x_j(t)$.

REMARK. If we simultaneously let $M\to\infty$, $N\to\infty$, then $X^{M,N}(.) \to \delta_{\mu(t)}$ where $\mu(.)$ satisfies the deterministic equation $\mu'(t) = L^{\mu(t)*}\mu(t)$.

The relevance of the two-level system to the optimization problem is that if $\rho > \rho_c$, $\alpha < 1$, then the level two interaction between search teams of strength $N^{-\alpha}$ is sufficient to direct individual search teams to the same global maximum. However this interaction is not sufficient to override a true difference between a global maximum and a local maximum unless it is also of order $N^{-\alpha}$. This means that the only stable local maxima x_ℓ (in the $M \to \infty$ limit) are those that satisfy $V(x_\ell) \geq V_{max} - KN^{-\alpha}$ for some constant K. In the next section we formulate as a conjecture a precise form of this statement and indicate the heuristic arguments which support it. These ideas motivate a rigorous study of large deviations and tunnelling for multilevel systems which is in progress.

5. HEURISTIC ANALYSIS OF THE MULTILEVEL SYSTEM

The equilibrium measures for the N-searcher process defined by (4.4) have densities of the form:

$$p_N(a;x_1,\ldots,x_N) = Z_N^{-1}\exp[\sigma^{-2}H(\bar{x})]\prod_{j=1}^{N}\rho(x_j) \tag{5.1}$$

where

$$\rho(x_j) = \exp\left(\sigma^{-2}[(1-\rho-\theta N^{-\alpha})x_j^2 - x_j^4]\right),$$

$$H(\bar{x}) = \tfrac{1}{2}N\rho(\bar{x})^2 + (\theta N^{1-\alpha}a + Nc)\bar{x}, \quad \bar{x} = N^{-1}\sum_{j=1}^{N} x_j,$$

and where a is obtained from the self-consistency equation

$$m(a + \theta^{-1}N^{\alpha}c) = a, \quad \text{and} \tag{5.2}$$

$$m(a) = \int\cdots\int x_1 p_N(a;x_1,\ldots,x_N)\,dx_1\ldots dx_N.$$

It follows that (cf. Dawson (1983)),

$$\left.\frac{dm(a)}{da}\right|_{a=0} = (N^{1-\alpha}\theta\sigma^{-2})v_N(0)$$

where $v_N(0) = \mathrm{VAR}(\bar{x}) = \mathrm{VAR}(\int xX(\infty,dx))$.

For $c = 0$ a necessary and sufficient condition for the existence of multiple equilibria is:

$$\left.\frac{dm(a)}{da}\right|_{a=0} > 1, \text{ i.e. } v_N(0) > (N^{1-\alpha}\theta\sigma^{-2})^{-1}. \tag{5.3}$$

It remains to determine the equilibrium variance $v_N(0)$ and it is this quantity that we obtain by a heuristic calculation.

Heuristically the N-searcher empirical process (<u>for large N</u>) satisfies an equation of the form:

$$\begin{aligned} dX_N(t) &= -\nabla F(X_N(t))dt + \sqrt{2/N}\,\sigma\nabla X_N(t)\cdot dW(t) \\ &\quad + \theta N^{-\alpha}\left([E[\textstyle\int yX_N(t,dy)) - \int yX_N(t,dy)]\nabla X_N(t)\right)dt \end{aligned} \tag{5.4}$$

where $F(.)$ is the action functional given by (3.7), ∇ denotes the infinite dimensional gradient and $W(.)$ is an infinite dimensional Wiener process. However if the system is close to the critical point $(\rho \approx \rho_c)$, then $m(t) = \int x X_N(t,dx)$ changes slowly compared to the other degrees of freedom and can be described (approximately) by the following stochastic differential equation:

$$dm(t) = -\nabla F(m(t))dt + \sqrt{2/N}\,\sigma dw(t) + N^{-\alpha}[a(t)-m(t)]dt \tag{5.5}$$

where $a(t) = E(m(t))$, $w(.)$ is a standard Brownian motion and

$$F(m) = \inf\{F(\nu) : \int x\nu(dx) = m\}.$$

If $\rho > \rho_c$, then it can be shown that F has global minima at $m_\ell = \int x\mu_\ell(dx)$ and $m_g = \int x\mu_g(dx)$ and a local maximum at $m_u = \int x\mu_u(dx)$. In addition $\partial^2F(x)/\partial x^2 > 0$ for $x = m_g$ or m_ℓ. If $\rho = \rho_c$ there is a unique equilibrium measure μ_u and in this case $\partial^2F(x)/\partial x^2|_{x=m_u} = 0$.

We now use the heuristic equation (5.5) to compute $v_N(0)$ in the various cases.

Case (i). $c = 0$ and $\rho < \rho_c$. In this case F has a unique minimum, $\partial^2F(x)/\partial x^2|_{x=m_u} > 0$, $v_N(0) = KN^{-1}$ and there is a unique invariant measure.

Case (ii). $\rho > \rho_c$.

(a) $c = 0$. In this case $\lim_{N\to\infty} v_N(0) = K$ and $\frac{dm(a)}{da}\Big|_{a=0} = KN^{1-\alpha}$ where K is a constant. Therefore if $0 \le \alpha < 1$ there is a pair of equilibrium measures concentrated near the two global maxima. On the other hand if $\alpha > 1$, there is a unique equilibrium measure.

(b) $c \neq 0$ but $N^\alpha|c|$ is sufficiently small. It can be shown that the conclusion of (a) remains valid in this case.

(c) $N^\alpha|c|$ large. In this case it follows from Equation (5.2) and properties of $m(.)$ that there exists a unique equilibrium near the global maximum of V.

Case (iii). $\rho = \rho_c$, $c = 0$. In this case $F(x) = K(x - m_u)^4$ near m_u, $v_N(0) = KN^{-\frac{1}{2}}$ and $\frac{dm(a)}{da}\Big|_{a=0} = K\theta N^{\frac{1}{2}-\alpha}$ where K is a constant. We conclude that there exist multiple equilibria either if $\alpha < \frac{1}{2}$ or if $\alpha = \frac{1}{2}$ and θ is sufficiently large. In the case $\alpha > \frac{1}{2}$ and $\rho = \rho_c$ there is a unique equilibrium measure.

REMARK. Let us underline the significance of the results of cases (i) b and c. They imply that in the $M \to \infty$ limit the system does not get trapped near a non-global local maximum unless it is *ε-optimal* where $\varepsilon = KN^{-\alpha}$ for some constant K, that is $V(x_\ell) \ge V_{max} - \varepsilon$. This suggests the possibility of designing a two level system which avoids being trapped near local maxima which are not ε-optimal. The exit time for escape from these non ε-optimal maxima is conjectured to be $O(e^{NQ_\ell})$. In general we can consider hierarchical search systems with successive sizes $N_1,\ldots,N_k$. With the appropriate choice of σ, $N_1, \ldots, N_k$ it may be possible to design a search system that will escape from all local maxima which are not ε_j-optimal (ε_j decreasing) with a time scale T_j associated with the jth level thus producing a system that gets to within a prescribed distance from the global maximum in a prescribed time scale.

6. CRITICAL POINT ANALYSIS

The heuristic arguments of the previous section suggest the possibility of using multilevel stochastic systems for global optimization and provides motivation for further study of these systems. As noted above the analysis simplifies near the critical point, $\rho = \rho_c$. In this section we give an indication of rigorous results for the critical fluctuations which provide partial confirmation for the conjectures produced above.

We begin with an analysis of the critical fluctuations as $N \to \infty$ for the non-linear $M_1(R^1)$-valued process X_N obtained in Proposition 4.1. Let Y_N denote the centered process with the central limit scaling and assuming that $X_N(0)$ is distributed according to the equilibrium distribution:

$$Y_N(t) = N^{\frac{1}{2}}[X_N(.) - \mu_u] \tag{6.1}$$

where μ_u is the unique invariant probability measure for X_N at $\rho = \rho_c$, that is,

$$L^{\mu_u *} \mu_u = 0.$$

The central limit theorem states that the processes $Y_N(.)$, when viewed as processes with values on a suitable space of generalized functions on R^1, converge weakly to a Gaussian process (cf. Hitsuda and Mitoma (1985), Mitoma (1985)). However at the critical point there exists a one-dimensional subspace in which the relevant fluctuations occur at a longer time scale and it is this time scale which yields insight into the questions raised above.

Consider the linearization L_u^* of the nonlinear operator $L^{\mu *}$ at $\mu = \mu_u$:

$$L^{\mu^*}[(1+v(.))\mu_u] \quad (L_u^* v)\mu_u + \text{higher order terms in v.}$$

The linearized operator L_u^* is given by

$$L_u^* v(x) = \sigma^2 \partial^2 v/\partial x^2 + ([(2-\rho)x-4x^3]\partial v(x)/\partial x) - \rho\left[\int y v(y)\mu_u(dy)\right]\left[\partial/\partial x\left(\frac{d\mu_u}{dx}\right)\right]. \tag{6.2}$$

Let $L_2^{\perp}(\mu_u)$ denote the subspace of $L_2(\mu_u)$ orthogonal to 1. We consider L_u^* acting on $L_2^{\perp}(\mu_u)$. At the critical point, $\rho = \rho_c$, the linearized operator L_u^* has a null vector $e_0(.) \in L_2^{\perp}(\mu_u)$. All other eigenvalues of L_u^* are strictly negative (cf. Dawson (1983)). The following theorem describes the fluctuations in the direction of e_0.

THEOREM 6.1. Let $\rho = \rho_c$ and μ_u, e_0 be as defined above. Assume that $X_N(0)$ is distributed according to the equilibrium distribution.

Case (i). $\alpha < \frac{1}{2}$. Define $Z_N(t) = N^{-\alpha/2} Y_N(N^{\alpha} t)$.

Then $Z_N(.) \to Z(.)$ in the sense of weak convergence of probability measures on $C([0,\infty), M_{\pm}(R^1))$ where $M_{\pm}(R^1)$ denotes the space of signed measures on R^1. The limit process has the form $Z(t) = z(t)e_0\mu_u$ where $z(.)$ is the equilibrium solution of the stochastic differential equation

$$dz(t) = -c_1 z(t) + \sigma_1 dw(t) \tag{6.3}$$

where $w(.)$ is a standard one-dimensional Brownian motion and $c_1, \sigma_1 > 0$.

Case (ii). $\alpha > \frac{1}{2}$. Define $Z_N(t) = N^{-1/4}\; Y_N(N^{1/2}t)$.

Then $Z_N(.) \to Z(.)$ in the above sense. The limit process $Z(.)$ has the form $Z(t) = z(t)e_0\mu_u$ where $z(.)$ is the equilibrium solution of the stochastic differential equation:

$$dz(t) = -c_2 z^3(t)dt + \sigma_2 dw(t), \tag{6.4}$$

where $c_2, \sigma_2 > 0$.

Idea of the Proof: The proof follows along the same lines as that of the single level case given in Dawson (1983). The generators of the rescaled processes $Z_N(.)$ are as follows:

Case (i). $\alpha < \frac{1}{2}$.

$$G_{Z_N} = N^{\alpha} G_1 + N^{(\alpha-\frac{1}{2})+\frac{1}{2}\alpha} G_2 + G_3 + N^{\frac{1}{2}(\alpha-1)} G_4 + G_5,$$

Case (ii). $\alpha > \frac{1}{2}$.

$$G_{Z_N} = N^{1/2}G_1 + N^{1/4}G_2 + G_3 + N^{-3/4}G_4 + N^{1/2-\alpha}G_5$$

where

$$L_u\phi = \sigma^2\partial^2\phi/\partial x^2 + [(2-\rho)x - 4x^3]\partial\phi/\partial x + \rho x\int \partial\phi/\partial y\,\mu_u(dy)$$

$$G_1F(\eta) = \textit{linearized evolution excluding } N^{-\alpha} \textit{ term}$$

$$= \int L_u\left(\frac{\delta F(\eta)}{\delta\eta(x)}\right)\eta(dx)$$

$$G_2F(\eta) = \textit{quadratic interaction term}$$

$$= \int\rho\int y\eta(dy)\left(\partial/\partial x\,\frac{\delta F(\eta)}{\delta\eta(x)}\right)\eta(dx)$$

$$G_3F(\eta) = \textit{second order operator (Gaussian noise)}$$

$$= \sigma^2\iint\left(\partial^2/\partial x\partial y\,\frac{\delta^2 F(\eta)}{\delta\eta(x)\delta\eta(y)}\right)\delta_x(dy)\mu_u(dx)$$

(where $\delta_x(dy)$ denotes unit mass at x)

$$G_4F(\eta) = \textit{lower order noise (negligable in this limit)}$$

$$= \sigma^2\iint\left[\partial^2/\partial x\partial y\ \frac{\delta^2 F(\eta)}{\delta\eta(x)\,\delta\eta(y)}\right]\delta_x(dy)\eta(dy)$$

$$G_5F(\eta) = \textit{level two mean-field feedback}$$

$$= -\theta\int\ \left(\partial/\partial x\ \frac{\delta F(\eta)}{\delta\eta(x)}\right)\int y\,\eta(dy)\ \mu_u(dx) + R(n)$$

In case (i) the terms which contribute to the limit are G_1, G_3, G_5 while in case (ii) they are G_1, G_3, G_2.

The argument proceeds by the perturbed test function method. For exmple in case (ii) we take

$$F_N = F_0 + N^{-1/4}F_1 + N^{-1/2}F_2$$

with $G_1F_0 = 0$. It can then be verified that

$$G_{Z_N}F_N = G_ZF_0 + R(N)$$

where $G_Z = [G_3 + G_2(G_1)^{-1}G_2]$ and $R(N)$ is a remainder term which disappears in the limit. We conclude that the limit process $Z(.)$ is characterized as a solution of the $M_{\pm}(R^1)$-valued martingale problem satisfying:

$$\left\{\frac{dZ(t)}{d\mu_u},\ \phi\right\} = 0 \quad \text{if } \{e_0,\phi\} = 0, \text{ and for each } f \in C_K^2(R^1),$$

$f\left(\left\{\frac{dZ(t)}{d\mu_0},e_0\right\}\right) - \int_0^t Af\left(\left\{\frac{dZ(s)}{d\mu_u},e_0\right\}\right)ds$ is a martingale where $\{.,.\}$ denotes the inner product in $L^2(\mu_u)$ and

$$Af(x) = \tfrac{1}{2}(\sigma_1)^2\partial^2 f(x)/\partial x^2 - c_1x^3\partial f(x)/\partial x \tag{6.5}$$

with $\sigma_1, c_1 > 0$. □

REMARK. This result confirms the conjecture based on the heuristic calculation that at $\rho = \rho_c$ the system (5.4) feels the effect of the level two mean field with coefficient $N^{-\alpha}$ before the stabilizing effect of $F(.)$ in the case $\alpha < \frac{1}{2}$ but not in the case $\alpha > \frac{1}{2}$.

7. MULTILEVEL CRITICAL POINT ANALYSIS

In this section multilevel fluctuations are considered at the critical point $\rho = \rho_c$ when $\alpha < \frac{1}{2}$ - this leads to an analysis involving three time scales. Let $X^{M,N}(.)$ denote the M_2-valued process defined by (4.1) and let μ_u, e_0 be defined as in Section 6. The centered process with central limit scaling is defined by

$$Y^{M,N}(t,\mu) = M^{1/2}[X^{M,N}(t,N^{1/2}\mu) - \delta_{\mu_0}(N^{1/2}\mu)]. \tag{7.1}$$

The appropriate class of test functions on M_2 for this process have the form:

$$F(\nu) = F_{f,h_1,h_2}(\nu)$$

$$= f\Big(\int_{M_1} h_1[\int_{R^1} h_2(x)\mu(dx)]\nu(d\mu)\Big)$$

where $h_2 \in C^2_K(R^1)$ and $h_1, f \in C^2(R^1)$. The scaling is implemented by setting

$$F_{f,h_1,h_2}(Y^{M,N}(.)) = F_{f,M^{\frac{1}{2}}h_1,N^{\frac{1}{2}}h_2}(X^{M,N}(.-\mu_u)).$$

The generator of the process $Y^{M,N}(.)$ has the form:

$$G_{Y^{M,N}}F(\nu) = (G_{1,1} + N^{-1/2}G_{1,2} + G_{1,3} + N^{-\alpha}G_{1,4} + M^{-1/2}N^{-(\frac{1}{2}+\alpha)}G_{2,1} + G_{2,2})F(\nu) \quad (7.2)$$

$$+ R(M,N,F)$$

for $\nu \in M_1(M_\pm)$ with $\nu\Big(\{\mu: \{\frac{d\mu}{d\mu_u}, 1\} = 0\}\Big) = 1$ where $\{.,.\}$ denotes the inner product in $L^2(\mu_u)$. $R(M,N,F)$ is a remainder term that disappears in the limit below. The operators $G_{i,j}$ are defined as follows: if $\int h_2(x)\mu_u(dx) = 0$,

$$G_{1,1}F(\nu) = \textit{linearized evolution excluding } N^{-\alpha} \textit{ term}$$

$$= f'(\nu)\int[h_1'(\mu)\int L_u h_2(x)\mu(dx)]\nu(d\mu)$$

$$G_{1,2}F(\nu) = \textit{quadratic interaction term}$$

$$= \rho f'(\nu)\int[h_1'(\mu)\int y\mu(dy)\int h_2'(x)\mu(dx)]\nu(d\mu),$$

$$G_{1,3}F(\nu) = \textit{level one noise}$$

$$= \sigma^2 f'(\nu)h_1''(\mu_u)\int[h_2'(x)]^2\mu_u(dx)$$

$$G_{1,4}F(\nu) = \textit{level two feedback to level one}$$

$$= \theta f'(\nu)\Big\{-\int[h_1'(\mu)\int y\mu(dy)\int h_2'(x)\mu_u(dx)]\nu(d\mu)$$

$$+ h_1'\Big(\int h_2(x)\mu_u(dx)\Big)\int[\int y\mu(dy)]\nu(d\mu)\int h_2'(x)\mu_u(dx)\Big\}$$

$$G_{2,1}F(\nu) = \textit{level two quadratic interaction}$$

$$= \theta f'(\nu)\int\Big(h_1'(\mu)\int h_2'(x)\mu(dx)\Big)\nu(d\mu)\int\int y\mu(dy)\nu(d\mu)$$

$$G_{2,2}F(\nu) = \textit{level two noise}$$

$$= \sigma^2 f''(\nu) h_1'(\mu_u)[\int h_2'(x)\mu_u(dx)]^2$$

We rescale the two level critical fluctuations in two stages:

(i) $V_1^{M,N}(t,\mu) = Y^{M,N}(N^{\alpha}t, N^{-\alpha/2}\mu)$ (cf. Theorem 6.1, (i))

(ii) $V_2^{M,N}(t,\mu) = (M^{1/2}N^{(1-\alpha)/2})^{-1/2}[V_1^{M,N}(M^{1/2}N^{(1-\alpha)/2}t,\mu) - M^{1/2}\nu_0]$

$$= (M^{1/2}N^{(1-\alpha)/2})^{-1/2}[Y^{M,N}(M^{1/2}N^{(1+\alpha)/2}t, N^{-\alpha/2}\mu) - M^{1/2}\nu_0]$$

where $\nu_0(\{\mu:\mu = ze_0\mu_u,\ a < z < b\}) = \int_a^b n(z)dz$

$$n(z) = (\sqrt{2\pi}\,\sigma_1)^{-1} \exp[\frac{-z^2}{2(\sigma_1)^2}]$$

and e_0 is defined as above.

THEOREM 7.1. Assume that $\rho = \rho_c$ and $\alpha < 1/2$ and $M \to \infty$, $N \to \infty$, $N^{(1-\alpha)}/M \to 0$, then $V_2^{M,N}(\cdot) \to V(\cdot)$ where $V(\cdot)$ is a $M_{\pm}(M_{\pm}(R^1))$-valued process satisfying

$$V(t,\{\mu:\mu = ze_0\mu_u,\ a<z<b\}) = \xi(t)\int_a^b [\partial n(z)/\partial z]dz$$

for $-\infty < a < b < \infty$, $\sigma_1, \sigma_2, v_2 > 0$ and $\xi(\cdot)$ satisfies the stochastic differential equation:

$$d\xi(t) = \sigma_2 dw(t) - c_2(\xi_t)^3 dt.$$

Idea of the Proof. The proof is carried out using the same ideas as those used in the proof of Theorem 6.1. In particular the generators associated with V_1 and V_2 are the following:

$$G_{V_1^{M,N}} = G_{2,0}^N + [M^{1/2}N^{(1-\alpha)/2}]^{-1} G_{2,1} + G_{2,2} + R(M,N)$$

where $R(M,N)$ is a remainder term that disappears in the limit below and

$$G_{2,0}^N = N^{\alpha}G_{1,1} + G_{1,3} + G_{1,4}, \text{ and}$$

$$G_{V_2^{M,N}} F_{f,h_1,h_2} = (K^{1/2}G_{2,0}^N + K^{1/4}G_{2,1} + G_{2,2})F_{f,h_1,h_2} + R(M,N)$$

where $K^{1/2} = M^{1/2}N^{(1-\alpha)/2}$, provided that $\int h_2(x)\mu_u(dx) = 0$ and $\int h_1[z\int h_2(y)e_0(y)\mu_u(dy)]n(z)dz = 0$.

Finally, the limiting generator is given by

$$G = \Pi[G_{2,2} + G_{2,1}(G_{2,0})^{-1}G_{2,1}],$$

where Π denotes the projection on the null space of $G_{2,0}$, and

$$G_{2,0} = \lim_{N\to\infty} G_{2,0}^N \quad \textit{(acting on null}(G_{1,1})\textit{)}.$$

The limit process lives on the set of ν such that

$$G_{2,0}F(\nu) = 0 \quad \text{for all} \quad F.$$

But this is the class of ν concentrated on signed measures of the form:

$$\nu(\{\mu:\mu = ze_0(.)\mu_u, a < z < b\}) = \int_a^b (n(z+c)-n(z))dz \qquad (7.3)$$

where c is an arbitrary real number. The final stage of the proof consists in identifying the limiting fluctuations on this subspace. This part is analogous to the proof ofTheorem 6.1 in case (ii). The limiting process $V(.)$ is characterized as the solution of a $M_\pm(M_\pm(R^1)$-valued martingale problem satisfying (7.3) and such that for $f \in C_b^2(R^1)$.

$$f\left(\int\left\{\frac{d\mu}{d\mu_u}, e_0\right\} V(t,d\mu)\right) - \int_0^t Af\left(\int\left\{\frac{d\mu}{d\mu_u}, e_0\right\} V(s,d\mu)\right)ds$$

is a martingale where A has the form (6.5). □

<u>*REMARKS*</u>. 1. Theorem 7.1 can be interpreted as follows:

$$X^{M,N}(M^{1/2}N^{(1+\alpha)/2}t, d\mu) \sim \delta_{\mu_u(1+xe_0)} \mathcal{N}(M,N,t;dx)$$

where $\mathcal{N}(M,N,t;dx)$ is a normal distribution on R^1 with

$$\text{mean} = M^{-1/4}N^{-(1-\alpha)/4}\xi(t) \qquad \text{and} \qquad \text{variance} = N^{-(1-\alpha)}(\sigma_1)^2.$$

2. The significance of this result is that for $\alpha < 1/2$, the interaction between search groups of strength $\theta N^{-\alpha}$ is sufficient to *slave* the individual search groups to their overall average. This means that in the case $\alpha < 1/2$ we can see at the level of the critical fluctuations the ability of the intergroup interaction to dominate the system thus providing evidence for the conjectures formulated in Section 5.

8. CONCLUSIONS FOR THE OPTIMIZATION PROBLEM

A one searcher system of the form (1.3) has the drawback that it can be trapped for times of order $O(e^{\Delta V/\sigma^2})$ near a local maximum of height ΔV. A N-particle search system of the form (3.1) can be trapped near local maxima only if ΔV exceeds some fixed value (which depends on σ and decreases as σ decreases). For local maxima of this type the system can be trapped for times of order $O(e^{NQ})$ where Q is a constant depending on σ. A two level system of the form (4.1) can be trapped near a local maxima only if it is ε-optimal where $\varepsilon = O(N^{-\alpha})$ for the longest natural time scale $O(e^{MN^{1-\alpha}Q})$. We conjecture that for a large class of functions V this system reaches an ε-optimal local maximum in the time scale $O(e^{N^{1+\alpha}}Q)$. This suggests the possibility of using hierarchically structured search systems to approach a global maximum in prescribed time scales.

Thus hierarchically structured parallel search systems appear to merit further study. However much more detailed analysis is required before any firm conclusions about the practical usefulness of this approach can be made. In addition there is need for consideration of other types of interaction. For example another natural interaction is one in which particles jump to the vicinity of the best values to have been discovered - this interaction will be studied elsewhere.

REFERENCES

E. Bonomi and J.L. Lutton (1984). The N-city travelling salesman problem: statistical mechanics and the Metropolis algorithm, SIAM Review 26, 551-568.

M. Cassandro, A. Galves, E. Olivieri, M.E. Vares (1984). Metastable behavior of stochastic dynamics: a pathwise approach, J. Stat. Phys. 35, 603-634.

F. Comets (1985). Tunnelling and nucleation for a local mean-field model, preprint.

V. Černý (1982). A thermodynamical approach to the travelling salesman problem: an efficient simulation algorithm. Preprint, Institut of Physics and Biophysics, Comenius Univ., Bratislava.

D. Dawson (1983). Critical dynamics and fluctuations for a mean-field model of cooperative behavior, J. Stat. Physics 31, 29-85.

D. Dawson (1985). Asymptotic analysis of multilevel stochastic systems, Lecture Notes in Control and Information Sciences 69, eds. M. Metivier and E. Pardoux, Springer-Verlag, 79-90.

D. Dawson and J. Gärtner (1986). Large deviations from the McKean-Vlasov limit for weakly interacting diffusions, Stochastics, to appear.

D. Dawson and J. Gärtner, in preparation.

C.C. Y. Dorea (1986). Limiting distribution for random optimization methods, SIAM J. Control, Vol. 24, 76-82.

M.I. Freidlin and A.D. Wentzell (1984). Random perturbations of Dynamical Systems, Springer-Verlag.

A. Galves, E. Olivieri, M.E. Vares (1984). Metastability for a class of dynamical systems subject to small random perturbations, preprint, I.H.E.S., Bures-sur-Yvette, France.

J. Gärtner (1985). Large deviations and tunnelling behavior of interacting particle systems, Abstract, Vilnius Conference.

J. Gärtner (1985). On the McKean-Vlasov limit for interacting diffusions, preprint.

S. Geman and C.R. Hwang (1985). Diffusions for global optimization, preprint.

M. Hitsuda and I. Mitoma (1985). Tightness problem and stochastic evolution equation arising from fluctuation phenomena for interacting diffusions, to appear.

S. Kirkpatrick, C.D. Gelatt and M.P. Vecchi (1983). Optimization by simulated annealing, Science 220, 671-680.

C. Léonard (1985). Une loi des grandes nombres pour des systèmes de diffusions avec interactions et à coefficients non bornés, LRSP Tech Rep. 48, Ottawa.

C. Léonard (1985). Large deviations and law of large numbers for a mean-field type infinite particle system, preprint.

H.P. McKean (1969). Propagation of chaos for a class of nonlinear parabolic equations, in *Lecture Series in Differential Equations*, Vol. 2, 41-57, Van Nostrand Reinhold, New York.

I. Mitoma (1985). An ∞-dimensional inhomogeneous Langevin equation, J. Fncl. Anal. 60.

K. Oëlschläger (1984). A martingale approach to the law of large numbers for weakly interacting stochastic processes, Ann. Prob. 12, 458-479.

M. Shiino (1985). H-theorem and stability analysis for mean-field models of non-equilibrium phase transitions in Stochastic Systems, Physics Letters Vol.112A, nos. 6,7, 302-306.

Y. Tamura (1984). On the asymptotic behavior of the solution of a nonlinear diffusion equation, J. Fac. Sci. Univ. Tokyo, Sect. 1A, Math. 31, 195-221.

Y. Tamura (1985). On the asymptotic behavior of the distribution of a diffusion process of the McKean type, 15th SPA, Nagoya, Japan.

H. Tanaka (1984). Limit theorems for certain diffusion processes with interaction, Proc. Taniguchi Symp. Katata and Kyoto, Japan, 1982, North Holland.

Remarks on the point interaction approximation.

R. Figari

Department of Theoretical Physics, University of Naples,
Naples, Italy.

G. Papanicolaou

Courant Institute, New York University, New York.

J. Rubinstein

Department of Mathematics, Stanford University,
Stanford, California.

1. Introduction.

The point interaction approximation is a way to study boundary value problems in regions with many small inclusions. For example, heat conduction in a material with many small holes that absorb heat, fluid flow in a region with many small obstacles, etc. The main idea is to replace the inclusions along with the boundary conditions on them by an inhomogeneous term in the differential equation which is then to hold evrywhere. As the name Point Interaction suggests, the effect of the inclusions is localized so the approximation is valid for small volume fractions. In this paper we shall consider the heat conduction problem as follows.

Let $u(x,t)$ be the temperature at position x in R^3 and at time $t \geq 0$. For $N=1,2,3,...$ let $w_1^N, w_2^N, ..., w_N^N$ be a sequence of points in R^3 and let D^N be the domain in R^3 defined by

$$D^N = \bigcap_{j=1}^{N} \left\{ x \mid \ |x - w_j^N| > \frac{\alpha}{N} \right\} \tag{1}$$

where α is a fixed positive constant. The temperature u satisfies the initial-boundary value problem

$$\frac{\partial u}{\partial t} = \Delta u \ \ in \ \ D^N \ , \ t>0 \tag{2}$$

$$u(x,0)=f(x) \ \text{ for } x \ in \ \ D^N \text{ and } \ u(x,t)=0 \ \ on \ \ \partial B_j^N \ \text{ for } \ t>0 \ , \ j=1,2,3,...,N \tag{3}$$

Here $f(x)$ is a smooth, positive function of compact support representing the initial temperature distribution and B_j^N is the sphere centered at w_j^N with radius α/N.

We are interested in the behavior of the solution u when N is large and the sequence of sphere centers w_j^N tends to a continuum. That is for every smooth function $\phi(x)$

$$\frac{1}{N}\sum_{j=1}^{N}\phi(w_j^N)\rightarrow\int V(x)\phi(x)dx \quad as \ N\rightarrow\infty \tag{4}$$

where $V(x)$ is the continuum sphere center density, assumed smooth and with compact support.

A related problem which is more realistic physically is the case of spherical inclusions that melt. This means that the radii of the spheres depend on time, are denoted by $\alpha_j^N(t)/N$ and we have the additional boundary condition

$$\frac{d\alpha_j^N(t)}{dt}=\frac{-1}{N}\frac{1}{4\pi\alpha_j^N(t)/N^2}\int_{|x-w_j^N|=\alpha_j^N(t)/N}\frac{\partial u(x,t)}{\partial n}dS(x)\ ,\ j=1,2,3,\cdots,N \tag{5}$$

Here n is the unit normal on the spheres pointing into the interior of the region D^N. At time zero the scaled radii are equal to $\alpha_0>0$

$$\alpha_j^N(0)=\alpha_0 \tag{6}$$

The boundary condition (5) is a simplified form of the usual one in free boundary problems: the rate of displacement of the boundary is proportional to the heat flux crossing the surface. It is simplified because the melting spheres do not change shape and their radii change in proportion to the average heat flux absorbed.

We will analyze here problem (2-4) in the continuum limit $N\rightarrow\infty$ by a relatively simple and direct method, the point interaction approximation. The radii of the spherical inclusions are already scaled in the above problems to be proportional to $1/N$. That this is appropriate scaling for a continuum limit can be seen easily by calculating the heat absorbed by a single sphere and requiring that N times this quantity be of order one as $N\rightarrow\infty$. The point interaction approximation is an intermediate step between (2-4) (or (2-6)) and the continuum limit which has features of both but is much simpler than (2-3) since the effect of the spheres is replaced by an appropriate point source term. The continuum approximation of (2-3) is the solution $\bar{u}$ of the initial value problem

$$\frac{\partial\bar{u}(x,t)}{\partial t}=\Delta\bar{u}(x,t)-4\pi\alpha V(x)\bar{u}(x,t) \quad x \ in \ R^3 \ ,\ t>0 \tag{7}$$

$$\bar{u}(x,0)=f(x)$$

Note that the volume fraction occupied by the speres goes to zero as $N\rightarrow\infty$ like N^{-2}. In the case of the melting spheres the continuum limit is given by the nonlinear diffusion equation

$$\frac{\partial\bar{u}(x,t)}{\partial t}=\Delta\bar{u}(x,t)-4\pi\alpha(x,t)V(x)\bar{u}(x,t) \quad x \ in \ R^3 \ ,\ t>0$$

$$\bar{u}(x,0)=f(x)$$

$$\frac{d\alpha(x,t)}{dt}=-\frac{\bar{u}(x,t)}{\alpha(x,t)}$$

$$\alpha(x,0)=\alpha_0$$

Note here the structure of the limit problem: it is a diffusion equation for the temperature field and an ordinary differential equation (a relaxation equation) for the continuum sphere radii. This is typical in problems where the point interaction approximation is called for as for example in waves in bubbly liquids [8].

The point interaction approximation for diffusion in a region with fixed spheres is described in section 2. A proof of its validity is given in the appendix.

Boundary value problems in regions with many small holes have been analyzed before in a variety of contexts and by several methods. Khruslov and Marchenko [1] use potential theoretic methods and give results in considerable generality regarding the possible distribution of the inclusion centers $\{w_j^N\}$, compatible with (4). Kac [2] studied (2), (3) when the points $\{w_j^N\}$ are independent identically distributed random variables over a region. He used properties of the Wiener sausage. Rauch and Taylor [3] formulated the results of Kac in a more analytic way and generalized them. Papanicolaou and Varadhan [4] studied (2-3) for nonrandom configurations of centers $\{w_j^N\}$ by probabilistic methods and obtained a strong form of convergence to the continuum limit. Ozawa [5] first considered the analysis of boundary value problems in regions with many small holes via a point interaction approximation. A study of the error in the continuum limit and a central limit theorem for it are given by Figari, Orlandi and Teta [6].

The point interaction approximation is a natural tool to analyze a variety of interesting problems in the continuum or homogenization limit. In the physical literature it goes back to Foldy's paper [7 see also 8,9] on sound propagation in a bubbly liquid and perhaps earlier. In almost all papers that followed Foldy's, the point interaction approximation is not treated as an important approximation in itself and averaging is carried out over the sphere center locations $\{w_j^N\}$. The closure problem that arises is then treated in a variety of ways depending on other parameters in the problem. In nonlinear cases, as with melting spheres and bubbles, the closure problems are much more involved. But averaging is not necessary. The continuum limit holds for deterministic sequences satisfying (4) and subject to some other conditions that hold for "most" realizations in the random case. The closure difficulties are thus avoided for many linear and nonlinear problems.

2. Point interaction approximation for diffusion in regions with many fixed inclusions.

We shall analyze the Laplace transform version of (1.2)

$$(-\Delta+\lambda)\, u^N(x) = f(x) \quad , \quad x \text{ in } D^N \;, \lambda > 0 \tag{1}$$

$$u^N(x) = 0 \quad , \quad | x - w_j^N | = \frac{\alpha}{N}$$

Let G be the free space Green's function

$$G(x,y) = \frac{e^{-\sqrt{\lambda}\, | x-y |}}{4\pi\, |x-y|} \tag{2}$$

Using Green's theorem we may rewrite (1) in integral form

$$u^N(x) = \int_{D^N} G(x,y) f(y)\, dy - \sum_{j=1}^{N} \int_{\partial B_j^N} G(x,y) \frac{\partial u^N(y)}{\partial n}\, dS(y) \tag{3}$$

where x is in D^N and n denotes the unit outward normal to the spheres ∂B_j^N .

Now let x tend to the surface of the ith sphere in (3). Using the Dirichlet boundary condition, we rewrite (3) in the form

$$\int_{\partial B_i^N} G(x,y) \frac{\partial u^N(y)}{\partial n}\, dS(y) + \sum_{\substack{j=1 \\ j\neq i}}^{N} \int_{\partial B_j^N} G(x,y) \frac{\partial u^N(y)}{\partial n}\, dS(y) \tag{4}$$

$$= \int_{D^N} G(x,y) f(y) dy$$

Let

$$\frac{1}{N} Q_j^N = \int_{\partial B_j^N} \frac{\partial u^N(y)}{\partial n}\, dS(y) \quad , \; j = 1,2,\ldots,N \tag{5}$$

be the charges induced on the spheres, suitably normalized. Since $f \geq 0$, the Q_j^N are nonnegative.

The spheres B_j^N have radius of order N^{-1} so they are small. We may then consider an approximate form of (4) where we place x at the center w_j^N of the ith sphere in the first term on the left and in the sum. We may also let the y in G in the sum in (4) go to the center w_j^N. Let us denote the approximate charges by q_j^N. Then

$$\frac{1}{4\pi\alpha} q_i^N + \frac{1}{N} \sum_{\substack{j=1 \\ j\neq i}}^{N} G(w_i^N, w_j^N)\, q_j^N = \int G(w_i^N, y) f(y) dy \qquad i = 1,2,\ldots,N \tag{6}$$

Note that we have also extended the integration on the right to all of R^3.

System (6) is what we call the point interaction approximation (PIA). The main point is to show that, under suitable conditions on the sequence of sphere centers $\{w_j^n\}$,

$$\lim_{N\to\infty} \sup_{1\le j\le N} | Q_j^N - q_j^N | = 0 \tag{7}$$

That is, the exact charges Q_j^N are well approximated by the the approximate carges q_j^N obtained from the PIA. Once this has been shown it is easy to pass to the continuum limit in (6) using the hypothesis (1.4).

Let $q(x)$ be the solution of the integral equation

$$\frac{1}{4\pi\alpha} q(x) + \int G(x,y)V(y)q(y)dy = \int G(x,y)f(y)dy \tag{8}$$

with x in R^3. If we define

$$u(x) = \frac{q(x)}{4\pi\alpha} \tag{9}$$

we see that (8) is the integral equation version of the Laplace transform of (1.7)

$$(-\Delta + \lambda)u(x) + 4\pi\alpha V(x)u(x) = f(x) \quad , \lambda > 0\,. \tag{10}$$

Using the regularity of the solution u of (10) and standard methods familiar from numerical analysis, along with hypothesis (1.4), we can show that

$$\lim_{N\to\infty} \sup_{1\le j\le N} | q_j^N - q(w_j^N) | = 0 \tag{11}$$

That is, (8) is the continuum limit of the PIA (6). Combining (7) and (11) we arrive at the desired convergence of the charges

$$\lim_{N\to\infty} \sup_{1\le j\le N} | Q_j^N - q(w_j^N) | = 0 \tag{12}$$

Once the charges have been shown to converge in the sense of (12) it is easy to show that $u^N(x)$, the solution of (1) converges to $u(x)$ the solution (10), outside a small set of points x near the surfaces of the spheres. We shall therefore concentrate here on proving (7) i.e. in proving the validity of the point interaction approximation.

3. Appendix.

For the proof we need some assumptions regarding the sequence of sphere centers $\{w_j^n\}$ in addition to (1.4). We will assume that

$$\inf_{i \neq j} | w_i^N - w_j^N | \geq \frac{1}{N^{1-\nu}} \quad \text{for } some \ \ 0 < \nu < \frac{1}{3} \tag{1}$$

$$\frac{1}{N^2} \sum_{\substack{i,j=1 \\ i \neq j}}^{N} \frac{1}{| w_i^N - w_j^N |^{3-\varepsilon}} \leq C \quad \text{for } some \quad \varepsilon > 0 \tag{2}$$

$$\frac{1}{N^3} \sum_{\substack{i,j,k=1 \\ i \neq j \neq k}}^{N} \frac{1}{| w_i^N - w_j^N |^2} \, \frac{1}{| w_k^N - w_i^N |^2} \leq C \tag{3}$$

These conditions are valid, for example, for sequences of independent identically distributed random sphere centers. They are valid in probability that is, there is a set of sphere center configurations that satisfy (1), (2) and (3) with probability arbitrarily close to one, uniformly in N.

The idea in introducing conditions like (1-3), contrary to what is done in [1] and [4], is that since they are valid in probability one need not worry about their form provided that they help to make the analytical part of the argument simple. This is important in nonlinear problems such as the melting spheres.

Let us now consider again the integral equation (2.4) where x is on ∂B_i^N. We average over the ith sphere

$$\frac{1}{4\pi(\frac{\alpha}{N})^2} \int_{\partial B_i^N} \int_{\partial B_i^N} G(x,y) \frac{\partial u^N(y)}{\partial n} \, dS(y) dS(x)$$

$$+ \sum_{\substack{j=i \\ j \neq i}}^{N} \frac{1}{4\pi(\frac{\alpha}{N})^2} \int_{\partial B_i^N} \int_{\partial B_j^N} G(x,y) \frac{\partial u^N(y)}{\partial n} \, dS(y) dS(x) \tag{4}$$

$$= \int_{\partial B_i^N} \int_{D^N} G(x,y) f(y) dy \, .$$

To simplify (4) we use the elementary identity.

$$\frac{1}{4\pi(\frac{\alpha}{N})^2} \int_{\partial B_i^N} G(x,y) dS(x) = \frac{1}{4\pi\, (\frac{\alpha}{N})} \, \frac{1 - e^{-2\sqrt{\lambda}\, \alpha/N}}{2\sqrt{\lambda}\, \alpha/N} \tag{5}$$

when y is on ∂B_i^N also. We use also the mean value theorem for any solution $h(x)$ of $(-\Delta + \lambda)h(x) = 0$

$$\frac{1}{4\pi r^2} \int_{|x|=r} h(x)dS(x) = \frac{\sinh\sqrt{\lambda}r}{\sqrt{\lambda}r} h(0) \tag{6}$$

Using (5) and (6), simplifying and recalling the definition (2.5) of the charges we obtain the following equations

$$\frac{\sqrt{\lambda}\,\alpha/N}{\sinh(\sqrt{\lambda}\,\alpha/N)} \frac{1-e^{-2\sqrt{\lambda}\alpha/N}}{2\sqrt{\lambda}\,\alpha/N} \frac{1}{4\pi\alpha} Q_i^N$$

$$+\frac{1}{N}\sum_{\substack{j=1\\ j\neq i}}^{N} G(w_i^N, w_j^N)\, Q_j^N = \int_{D^N} G(w_i^N, y) f(y)dy + \eta_i^N , \tag{7}$$

$$i = 1,2,\ldots,N$$

where

$$\eta_i^N = \sum_{\substack{j=1\\ j\neq i}}^{N} \int_{\partial B_j^N} \left[G(w_i^N, y) - G(w_i^N, w_j^N) \right] \frac{\partial u^N(y)}{\partial n} dS(y) . \tag{8}$$

Lemma 1.

$$\sup_{1\leq i\leq N} |\, \eta_i^N \,| \leq \frac{C}{N^{1+\nu}} \sum_{\substack{j=1\\ j\neq i}}^{N} G(w_i^N, w_j^N)\, Q_j^N . \tag{9}$$

Proof: We simply note that for y on ∂B_j^N

$$\frac{e^{-\sqrt{\lambda}\alpha/N}\, |\, w_i^N - w_j^N \,|}{|\, w_i^N - w_j^N \,| + \frac{\alpha}{N}} \leq \frac{G(w_i^N, y)}{G(w_i^N, w_j^N)} \leq \frac{e^{\sqrt{\lambda}\alpha/N}\, |\, w_i^N - w_j^N \,|}{|\, w_i^N - w_j^N \,| - \frac{\alpha}{N}}$$

From this and hypethesis (1) the estimate (9) follows easily.

Now we return to equations (7) and use the estimate (9) along with the positivity of the Q_j^N . We see that for N large the error η_i^N can be absorbed into the sum on the left of (7) without changing signs. This and the fact that $f \in C_0^\infty(R^3)$ give us the a-priori estimate

$$0 \leq Q_i^N \leq C \tag{10}$$

where all constants C that appear (possibly different constants) are independent of N. We also have the estimate

$$0 \leq \frac{1}{N} \sum_{\substack{j=1\\ j\neq i}}^{N} G(w_i^N, w_j^N)\, Q_j^N \leq C . \tag{11}$$

Combining (10) and (9) we have now an estimate for the error

$$\sup_{1 \le i \le N} |\eta_i^N| \le \frac{C}{N^\nu} \qquad \nu > 0 . \tag{12}$$

If we also denote by $a_N(\lambda)$ ($\lambda > 0$ fixed) the coefficient in front of $\frac{1}{4\pi\alpha} Q_i^N$ in (7) then clearly

$$a_N \to 1 \quad \text{as} \quad N \to \infty \tag{13}$$

Put

$$f_i^N = 4\pi\alpha \int_{D^N} G(w_i^N, y)dy \quad , \quad i = 1,2,\ldots,N \tag{14}$$

and let q_i^N satisfy the PIA (2.6). We write this system in the form

$$(I + A^N)q^N = f^N \tag{15}$$

where $I = (\delta_{ij})$ is the identity matrix and

$$A_{ij}^N = \frac{4\pi\alpha}{N} G(w_i^N, w_j^N) \quad i \ne j \tag{16}$$

$$A_{ij}^N = 0 \qquad i = j$$

The vectors q^N and f^N have components q_i^N and f_i^N. Let $|| \;\; ||$ stand for the l^2 vector (or matrix) norm on R^N.

Lemma 2. For $\lambda > 0$ large enough.

$$\sup_N || A^N || < 1 \tag{17}$$

Proof. Clearly

$$|| A^N ||^2 \le \frac{4\pi\alpha}{N^2} \sum_{\substack{i,j=1 \\ i \ne j}}^{N} G(w_i^N, w_j^N)^2 \tag{18}$$

and this is controlled by (3) and (1.4). In addition, we can use the exponential factor in G that has the parameter λ and we can make the right side of (18) small by choosing λ large.

Thus, (15) is invertible uniformly in N and we may write

$$q^N = (I + A^N)^{-1} f^N \tag{19}$$

Clearly

$$0 \le q_i^N \le C \quad , \quad i = 1, 2, ..., N \tag{20}$$

and since the f_i^N are also uniformly bounded we have that

$$\sum_{j=i}^{N} \left[\frac{1}{\sqrt{N}} f_i^n \right]^2 \le C \tag{21}$$

Hence from (17) we deduce that

$$\frac{1}{N} \sum_{j=i}^{N} (q_i^N)^2 \le C \tag{22}$$

To estimate the difference between Q_j^N and q_j^N we need an additional estimate on the errors η_i^N .

Lemma 3.

$$\sum_{i=1}^{N} | \eta_i^N |^2 \le \frac{C}{N} \tag{23}$$

Proof. From (8) and (10) we see that

$$| \eta_i^N | \le \frac{C}{N^2} \sum_{\substack{j=1 \\ j \ne i}} \frac{1}{| w_i^N - w_j^N |^2}$$

We note that this does not follow from the estimate (9). One has to consider the η_i^N from (8) directly. From (2.4) we get

$$\sum_{i=1}^{N} | \eta_i^N |^2 \le \frac{C}{N^4} \sum_{i=1}^{N} \sum_{\substack{j \ne 1 \\ j \ne i}}^{N} \sum_{\substack{k=1 \\ k=i}}^{N} \frac{1}{| w_j^N - w_i^N |^2} \frac{1}{| w_k^N - w_i^N |^2}$$

$$= \frac{C}{N} \left\{ \frac{1}{N^3} \sum_{\substack{i,j,k=1 \\ j \ne i \ne k}}^{N} \frac{1}{| w_i^N - w_j^N |^2} \frac{1}{| w_k^N - w_i^N |^2} \right.$$

$$\left. + \frac{1}{N^3} \sum_{\substack{j,k=1 \\ j \ne k}}^{N} \frac{1}{| w_j^N - w_k^N |^4} \right\}$$

The first sum on the right is controled by hypothesis (3). In the second sum we note that because of hypothesis (1) and (2)

$$\frac{1}{N^3} \sum_{\substack{j,k=1 \\ j \ne k}}^{N} \frac{1}{| w_i^N - w_k^N |^4} \le C \frac{(N^{1-\nu})^{1+\varepsilon}}{N} \frac{1}{N^2} \sum_{\substack{j,k=1 \\ j \ne k}}^{N} \frac{1}{| w_j^N - w_k^N |^{3-\varepsilon}} \le C$$

This estimate completes the proof of the lemma.

We can now estimate the difference $| Q_j^N - q_i^N |$ between the exact charges and the approximate charges of the PIA. In fact by subtracting (15) from (7) we see that

$$(I + A^N)(Q^N - q^N) = (1 - a^N)Q^N + \eta^N \tag{25}$$

Now $| 1 - a^N | \le C/N$ for λ fixed and $0 \le Q_i^N \le C$. Thus the l^2 norm of the right side of (25) is less than a constant times $N^{-1/2}$, in view of lemma 3. From lemma 2 and this observation we conclude that

$$\sum_{j=1}^{N} (Q_j^N - q_j^N)^2 \le \frac{C}{N} \tag{26}$$

This implies that

$$\sup_{1 \le j \le N} | Q_j^N - q_j^N | \le \frac{C}{N^{1/2}} \tag{27}$$

We summarize the above in the following theorem.

Theorem

Under hypotheses (1), (2), (3) and (1.4) the charges Q_j^N of problem (2.1), defined by (2.5), and the approximate charges q_j^N satisfying the PIA (2.6) (or (15)) are uniformly close as $N \to \infty$ in the sense of (27).

As we indicated in section 2, the elementary estimate (2.11) gives us the validity of the continuum approximation (2.12) for the charges and hence for the full solution of (2.1) and (2.10).

Acknowledgment

The work of George Papanicolaou was supported by the National Science Foundation and the Office of Naval Research. The work of Rodolfo Figari was partially supported by the Office of Naval Research. A slightly different version of this paper appeared in the Springer Lecture Notes in Biomathematics, vol. 70 (1987), pp. 202-209.

References

[1] E. I. Khruslov and V. A. Marchenko, Boundary value problems in regions with fine-grained boundaries, Naukova Dumka, Kiev, 1974.

[2] M. Kac, Probabilistic methods in some problems of scattering theory, Rocky Mountain J. Math. 4, 1974, 511-538.

[3] J. Rauch and M. Taylor, Potential and scattering theory on wildly perturbed domains, J. Funct. Anal. 18, 1975, 27-59

[4] G. Papanicolaou and S.R.S. Varadhan, Diffusion in regions with many small holes. In Stochastic Differential Systems (ed. B. Grigeliouis). Lecture Notes in Control and Information Theory 25, 190-206, Springer.

[5] S. Ozawa., On an elaboration of M. Kac's Theorem concerning eigenvalues of the Laplacian in a region with randomly distributed small obstacles, Comm. Math. Phys. 91, 1983, 473-487.

[6] R. Figari, E. Orlandi and J. Teta, The Laplacian in regions with many small obstacles: fluctuations around the limit operator 41, 1985, 465-488.

[7] L. L. Foldy, The multiple scattering of waves, Phys. Rev. 67, 1945, 107-119.

[8] R. Caflisch, M. Miksis, G. Papanicolaou and L. Ting, Effective equations for wave propagation in bubbly liquids, J. Fluid Mech. 153, 1985, 259-273 and also 160, 1-14.

[9] J. Rubinstein, NYU Dissertation, 1985.

RENORMALIZATION GROUP TREATMENT OF THE HYDRODYNAMICS OF POLYMER CHAINS IN THE RIGID BODY APPROXIMATION

Karl F. Freed

Shi-Qing Wang

Jack F. Douglas

The James Franck Institute and Departments of Chemistry and Physics

The University of Chicago
Chicago, Illinois 60637

Abstract

The Kirkwood-Riseman model is solved using renormalization group methods for the hydrodynamic radius R_H through order ε^2 for random walk (Gaussian) chains where $\varepsilon=4-d$ and d is the dimensionality of space. These calculations are used, in part, to investigate the effects of the preaveraging approximation within the rigid body hydrodynamic interaction model. Our calculations produce a preaveraging correction of roughly double that found in Monte Carlo simulations, indicating the need for further higher order calculations and/or refinements of the dynamical renormalization group method. We also consider the crossover dependence of R_H on the strength of hydrodynamic interactions for a Gaussian chain, showing that our results closely coincide with older classical descriptions of the hydrodynamic radius. For example, our preaveraging Kirkwood-Riseman solution is found to be identical to the Kirkwood approximation to R_H to order ε^2. Preaveraging affects the crossover dependence in order ε^2.

I. Introduction

The mathematical description of the dynamics of a single polymer molecule in solution involves the complicated problem of the hydrodynamics of strongly interacting particles. To a first approximation, a polymer in solution may be modeled as a set of sequentially bonded monomer units in a continuous fluid. Such a description of polymers in solution ignores the molecular nature of the solvent as well as the internal molecular structure of the monomer itself and is only valid for the calculation of large length scale polymer properties which depend on distances large compared to monomer or solvent molecular dimensions. Particular long wavelength polymer properties of this nature are the polymer translational diffusion coefficient and the intrinsic viscosity.

Discrete bead models [1] of the dynamics of polymers in solution take each of the sequentially bonded monomers to be point sources of frictional resistance to the solvent fluid flow. The spatial distribution, or conformation, of the polymer is that of a random walk with perhaps interactions between nonbonded monomers (called excluded volume interactions). If $\underset{\sim}{R}_i(t)$ designates the position of the ith monomer at time t with respect to an arbitrary spatial origin, then the conformation $\{\underset{\sim}{R}_i(t)\}$ of the chain at any time is an extremely irregular "fractal" object. The hydrodynamics of rigid suspended particles [2], on the other hand, leads to solvable mathematical problems for particles of regular shapes such as spheres, ellipsoids, cylinders, etc., but not for particles having highly irregular geometries like those of random coil polymers. These difficulties appear to be compounded by the fact that the general principles of statistical mechanics dictate that our interest lies not just in the dynamics of a single polymer with conformation $\{\underset{\sim}{R}_i(t)\}$ but in the ensemble average of polymer properties over all possible conformations $\{\underset{\sim}{R}_i(t)\}$. This latter feature is a blessing because the average properties are considerably simpler than those of any particular member of the ensemble.

The great complexity of the mathematical treatment of the dynamics of polymers in solution has led to the development of a hierarchy of different polymer models of increasing sophistication. Perhaps the simplest, retaining the chain-like structure of the polymer, is the model due to Kirkwood and Riseman [3] in which the polymer is assumed to be rigidly frozen in one of a large number of possible conformations. Even this highly idealistic picture of polymer dynamics is analytically intractable, in general, and has required further approximation.

Kirkwood and Riseman [3,4] introduce an additional "preaveraging" approximation in which the chain conformation dependent dydrodynamic disturbance, induced by the motion of the polymer on the fluid flow, is replaced by its equilibrum average over all chain conformations, an approximation which is in the same physical spirit as the average introduced by Hartree to describe the motion of elections in an atom. The preaveraging approximation introduces errors of both mathematical and physical nature. The analytical treatment of the corrections to the preaveraging aproximation has remained as a fundamental theoretical challenge in polymer science. While the preaveraged Kirkwood-Riseman theory [3] is successful in explaining many dynamical properties of real (nonrigid) polymers, recent Monte Carlo simulations [5,6] using Kirkwood-Riseman theory and random walk models of rigid polymers indicate the presence of a considerable preaveraging corrections for flexible polymers in the Kirkwood-Riseman rigid body approximation.

Our analysis of the preaveraging corrections within the Kirkwood-Riseman model proceeds using renormalization group methods [7-9] which consider the analytic continuation of the Kirkwood-Riseman model to continuous spatial dimensionality d. Despite the strongly interacting dydrodynamic nature of the polymer dynamics, renormalization group methods utilize perturbation expansions in the small parameter ε=4-d as explained below. The analytic continuation to ε of order unity is performed by considering the singularities obtained in the formal mathematical limit of d $\rightarrow$ 4. The information contained in these singularities is used to resum [8,9] the ε-expansion into a form which is valid for describing strongly interacting systems. The details of the renormalization group method are rather lengthy [8,9], so only a brief sketch of some of its features is given here in order to focus upon some of the basic results and concepts of the renormalization group theory.

The general Kirkwood-Riseman renormalization group method has been developed by Oono and Kohmoto [8] who have limited their calculations to the first order in ε because of the Kirkwood-Riseman model neglects [9,10,11] contributions which occur in a physically more precise description from solvent velocity fluctuations and which may begin to contribute to the polymer diffusion coefficient and intrinsic viscosity in order ε^2. However, the availability of Monte Carlo simulations of polymer dynamics in the rigid body approximation provides a useful benchmark against which to test the dynamical renormalization group theory in a completely unambiguous manner. Furthermore, preaveraging corrections for the polymer diffusion coefficient begin [12] only in order ε^2, so these technically complicated order ε^2 calculations are necessary to theoretically treat this long standing problem. [12] We note in passing that a range of dynamical models of polymers [10,11] produce identical descriptions of the hydrodynamic radius to order ε, and thus the dynamical theory to this order in ε is quantitatively rather crude.

When excluded volume [13] is absent, polymers under theta conditions behave to first order as if they have the conformations of random walks with no polymer-polymer interactions. The dynamics of polymers in theta conditions can be reproduced by use of a simple model which takes the polymer to be a hard sphere of radius given by the radius of gyration of the polymer. [14] This model views the fluid within the polymer coil to be entrapped within it and not to freely exchange or "drain" through the coil. Thus, polymer dynamics of theta solution polymers is governed by the nondraining limit of the hydrodynamic theory.

When very repulsive polymer-polymer interactions are present, the polymers take up the conformations of self-avoiding random walks. [14] Our previous calculations [15,16] of the polymer diffusion coefficient and intrinsic viscosity, using renormalization group methods applied to approximate d=3 analytical solution to the Kirkwood-Riseman equations, show that, as the polymer begins to expand due to the excluded volume interactions, the fluid has the opportunity to stream through the outer periphery of the polymer coil. Hence, draining may contribute to the polymer dynamics in good solvents, and this may lead to diminished dynamical measures of the polymer's dimensions compared to that given by static measurements such as the radius of gyration or the second virial coefficients. This theoretical result explains why dynamical measures of polymer dimensions do not vary with excluded volume in the same fashion as do equilibrium measures. It is, therefore, of great interest to also consider the role of excluded volume within the dynamical renormalization group Kirkwood-Riseman formulation [17] to determine whether draining plays a central role in good solutions and whether it enters qualitatively in the same fashion as in our previous more approximate treatment [16]. While the role of excluded voume on polymer dynamics is both of theoretical and practical interest, we limit the discussion here to the case of random walk (Gaussian) chains under theta conditions in order to illustrate the application of renormalization group methods to polymer dynamics. Contributions from excluded volume are described in Refs. 12 and 17.

It should be emphasized that the approximate nature of the Kirkwood-Riseman model implies that the problems must eventually be reconsidered from the viewpoint of more realistic models of the dynamics of flexible polymer chains. Nevertheless, the Kirkwood-Riseman treatment is expected to provide many of the theoretical methods and insights which will be of use in such further studies.

II. Kirkwood-Riseman Theory

The actual computations are performed using continuous random walk models of the polymer chain confrontation [18] as this continuum limit does not affect the long wavelenght properties of interest here. Within this continuous representation, let $\underset{\sim}{F}[\underset{\sim}{R}(x)]$ be the friction force exerted on the chain segment at the spatial position $\underset{\sim}{R}(x)$ where x designates a contour variable, running between 0 and N along the chain, and where N is the chain length. Then, the friction coefficient f is defined in terms of the equilibrium average of this friction force as

$$f\underset{\sim}{v}_d = \ell^{-1}\int_0^N dx \langle \underset{\sim}{F}[\underset{\sim}{R}(x)]\rangle , \quad (1)$$

where ℓ is the length of a monomer unit, the average is over all chain conformations and $\underset{\sim}{v}_d$ is the drift velocity of the whole chain. The polymer chain in the Kirkwood-Riseman model is taken to be rigid, so every chain segment has the same drift velocity $\underset{\sim}{v}_d$.

Hydrodynamic interactions between the point friction centers are represented in terms of the usual [2,18] Oseen tensor $\underset{\approx}{T}[\underset{\sim}{R}(x),\underset{\sim}{R}(x')]$. Solution of the steady state linearized Navier-Stokes equations for a set of point friction sources gives the familiar force balance equation of Kirkwood and Riseman [3,4],

$$\underset{\sim}{F}[\underset{\sim}{R}(x)] = \zeta_0\underset{\sim}{v}_d-(\zeta_0/\ell)\int_0^N dx'\underset{\approx}{T}[\underset{\sim}{R}(x),\underset{\sim}{R}(x')]\cdot\underset{\sim}{F}[\underset{\sim}{R}(x')], \quad (2)$$

for stationary fluid flow in the rigid body approximation where ζ_0 is the bead friction coefficient. Equation (2) is complicated by virtue of the fact that the chain conformation $\underset{\sim}{R}(x)$ is that of a random walk and that the Oseen tenson $\underset{\approx}{T}[\underset{\sim}{R}(x),\underset{\sim}{R}(x')]$ explicitly depends on $\underset{\sim}{R}(x)$ (see below). Some simplification ensues because equation (1) requires only the average force which from (2) obeys the equation

$$\langle \underset{\sim}{F}[\underset{\sim}{R}(x)]\rangle = \zeta_0 \underset{\sim}{v}_d - (\zeta_0/\ell)\int_0^N dx' \langle \underset{\approx}{T}[\underset{\sim}{R}(x),\underset{\sim}{R}(x')]\cdot \underset{\sim}{F}[\underset{\sim}{R}(x')]\rangle , \tag{3}$$

but equation (3) is not closed.

Preaveraging, on the other hand, replaces (3) by the preaveraged Kirkwood-Riseman equation

$$\langle \underset{\sim}{F}[\underset{\sim}{R}(x)]\rangle_p = \zeta_0 \underset{\sim}{v}_d - (\zeta_0/\ell)\int_0^N dx' \langle \underset{\approx}{T}[\underset{\sim}{R}(x),\underset{\sim}{R}(x')]\rangle \cdot \langle \underset{\sim}{F}[\underset{\sim}{R}(x')]\rangle_p , \tag{4}$$

where the subscript p denotes preaveraging. Notice that the average $\langle \underset{\approx}{T}[\underset{\sim}{R}(x),\underset{\sim}{R}(x')]\rangle$ depends only on x and x', but not on the chain conformation $\underset{\sim}{R}(x)$, so (4) is a closed equation whereas (3) is not. Approximate solutions to (4) have been extensively studied and are reviewd by Yamakawa [4], while (3) remains to be solved. The difference between the solution of (1) and (3) and that of (1) and (4) gives the rigid chain preaveraging error we seek.

III. Analytic Continuation and Renormalization Group

We desire the solution to (3) for random walk polymers in three-dimensions, but it proves convenient [9] to study the problem for arbitrary continuous dimensionality d. This analytic continuation to continuous d is defined through the random walk configurations in d-space and the d-space form of the Oseen tensor which results from the d-space Navier-Stokes equation in Fourier form [8,10,18]

$$\underset{\approx}{T}_{\alpha\beta}[\underset{\sim}{R}(x),\underset{\sim}{R}(x')] = (2\pi)^{-d}\int d^dk\,(k^2\eta_s)^{-1}(\delta_{\alpha\beta} - k^{-2}k_\alpha k_\beta)\exp\{i\underset{\sim}{k}\cdot[\underset{\sim}{R}(x) - \underset{\sim}{R}(x')]\} , \tag{5}$$

where α and β denote the Cartesian components of vectors and tensors and η_s is the solvent viscosity.

Because the random walk polymer dimensions $\underset{\sim}{R}(x)$ scale as the radius of gyration which is proportional to $N^{1/2}$ dimensional analysis of (3) and (5) shows that the quantity (ζ_0/n_s) in (3) enters only in the dimensionless combination of $(\zeta_0/n_s)(N/\ell)^{\varepsilon/2}$ where $\varepsilon = 4 - d$. Thus, attempts at iteratively solving (3) with expansions in powers of the Oseen tensor (5) are of little use for $N \to \infty$ and $d < 4$. For $d > 4$ ($\varepsilon < 0$) the hydrodynamic interactions become a small perturbation and lead for a long polymer chains to the free-draining solutions $\underset{\sim}{F}[\underset{\sim}{R}(x)] = \zeta_0 \underset{\sim}{v}_d$. This arises because space is so open for $d > 4$ that fluid streamlines easily pass through the interior of a random walk polymer. At the borderline or "critical" dimensionality $d = 4$, where the parameter (ζ_0/n_s) is dimensionless, expansions in (ζ_0/n_s) are permitted for small values of this parameter.

However, we desire solutions of the Kirkwood-Riseman equations for $d = 3$, not for $d = 4$. The renormalization group methods perturbatively study the Kirkwood-Riseman equations near four dimensions, where perturbation theory is reasonable, by utilizing expansions in ε along with those in (ζ_0/n_s) and by analytically continuing in ε based on the singularities present in these expansions. These expansions are probably asymptotic for the full model (without the preaveraging approximation), but may become absolutely convergent upon use of the mean field like preaveraging approximation.

The origin of these singularities may be deduced from the structure of (3) and of the d-space Oseen tensor (5). Averages of (5) or of products of Oseen tensors have singularities from the $x = x'$ contributions which are to be excluded from calculations since they correspond to self-hydrodynamic interactions which are already included in the friction coefficients ζ_0. This may be accomplished by introducing a cut-off on the x-integrals such that integration limits have the minimum distance cut-off $|x-x'| > a$, a result which naturally follows in passing to the continuum limit of the discrete chain Kirkwood-Riseman model. Alternatively, these self-interaction contributions are discarded in the continuum representation (3) as follows: When $d < 2$, these integrals can formally be shown to converge, so the nonsingular integrals of interest may be obtained by the method of dimensional regularization [8,9] in which the integrals are evaluated for $d < 2$ and $a \to 0$ and then are analytically continued to $d < 4$. [19] When this procedure is applied, it is found that the perturbation expansion in ε and (ζ_0/n_s) develops poles in ε. As the original theory should be well defined for $d = 4$ or $\varepsilon = 0$, the renormalization group provides a redefinition (called a renormalization) of the expansion such that the poles in ε are removed through the definition of renormalized variables.

IV. Perturbation Calculations and Renormalization

The iterative solution [3,8] of (2) as a series in $\underset{\approx}{T}$ is substituted into (1) where the avaraging takes the polymer to have the conformation of a random walk. Terms through order $(\underset{\approx}{T})^2$ are retained, and an expansion is made in powers of ε. The final results are obtaind as [12]

$$f = F_0 z_H^0\{1 + (2z_H^0/\varepsilon) + (2z_H^0/\varepsilon)^2(1 + 7\varepsilon/24) + O[(z_H^0)^3]\}, \quad z_H^0 \ll 1, \tag{6}$$

where the convenient dimensionless hydrodynaic variable is written as [12]

$$z_H^0 = (d/2\pi\ell^2)^{(d-2)/2}(\zeta_0/n_s)[2(d-1)/d\pi(d-2)(6-d)](N/\ell)^{\varepsilon/2} , \tag{7}$$

and where the prefactor in (6) for random walk polymers is given by [12]

$$F_0 = [d^2(d-2)(6-d)(2\pi/d)^{d/2}(N\ell)^{(d-2)/2}n_s]/4(d-1). \tag{8}$$

The perturbation expansion in z_H^0 has a formal poles in ε as mentioned above, and these are to be removed by renormalization.

The renormalization group effectively introduces an averaging or coarse graining over lengths L. This averaging over lengths L leaves large length scale polymer properties like the friction coefficient f unchanged. The coarse grained hydrodynamic length scale L appears in the definition of the bare dimensionless hydrodynamic coupling constant u_H^0 which is defined through

$$z_H^0 \equiv u_H^0(2\pi N/L)^{\varepsilon/2} . \tag{9}$$

The expansion (6) has z_H^0 as a natural dimensionless variable, but the renormalization group method generally requires an N-independent dimensionless variable such as u_H^0. Because the friction coefficient f should exist when $\varepsilon \to 0$, the renormalized variable u_H is defined in terms of u_H^0 and of the renormalization constant $Z_{H\theta}$ through

$$u_H = Z_{H\theta} u_H^0 \tag{10}$$

and through the "renormalizability" condition that $Z_{H\theta}$ be determined as an expansion in u_H such that the resultant expansion of f (or any other dynamical polymer property) in terms of the renormalized variable u_H is finite as $\varepsilon \to 0^+$. Thus, $Z_{H\theta}$ is determined as [12]

$$Z_{H\theta} = 1 + (2u_H/\varepsilon) + (2u_H/\varepsilon)^2 (1 - 7\varepsilon/24) + O[(u_H)^3] , \tag{11}$$

and the renormalized perturbation expansion for f becomes

$$f = F_0 u_H (2\pi N/L)^{\varepsilon/2} [1 - u_H \ln(2\pi N/L)] + O[(u_H)^3] , \tag{12}$$

where the prefactor in (12) has been written using the expansion $(2\pi N/L)^{\varepsilon/2} = 1 + (\varepsilon/2)\ln(2\pi N/L) + O(\varepsilon^2)$. Equation (12) is evidently well defined as $\varepsilon \to 0$ as required; however, this equation has still only been derived under the assumptions that ε and u_H are small, while we desire solutions when ε and u_H are of order unity. This analytic continuation out of the perturbative regime is obtained from the solution of the renormalization group equation in the next section.

We conclude this section by quoting the final results obtained by pursuing the same type of analysis with the preaveraged approximation (4). Letting sub or superscripts p designate preaveraged quantitites, we find [12]

$$f_p = F_0 z_H^0 \{1 + (2z_H^0/\varepsilon)^2 + (2z_H^0/\varepsilon)^2 + O[(z_H^0)^3]\} , \tag{13a}$$

$$f_p = F_0 u_{Hp} (2\pi N/L)^{\varepsilon/2} [1 - u_{Hp} \ln(2\pi N/L)] + O[(u_{Hp})^3] , \tag{13b}$$

$$Z_{H\theta}^{(p)} = 1 + (2u_{Hp}/\varepsilon) + (2u_{Hp}/\varepsilon)^2 + O[(u_H)^3] . \tag{14}$$

Note that equations (12) and (13b) are <u>identical</u> apart from the difference in the renormalized hydrodynamic variables which arises because of the differences in the renormalization constants in (11) and (14).

Renormalization is introduced here as a mathematical device to ensure that the the renormalized perturbation expansions in u_H and ε exist for $\varepsilon \rightarrow 0$, but no proof is given where that this procedure can be accomplished to all orders in ε. Such a proof of the renormalizability of the theory requires a consideration of the nature of all the singularities in higher orders and the demonstration that these singularities may be removed using a finite number of renormalization constants. [20]

V. Renormalization Group Equation

Consider a generic dimensionless large length scale hydrodynamic quantity H, such as f/F_0, which is calculated within the Kirkwood-Riseman model as a function of the bare parameter ζ_0/n_s and of N. We represent this scaling function in the form

$$H_B = H_B(\zeta_0/n_s, N) \quad . \tag{15}$$

Taking the derivative of H_B with respect to L with the arguments of H_B held constant gives by definition

$$L(\partial/\partial L)H_B|_F \equiv 0 \ , \tag{16}$$

where the subscript F indicates that the variables ζ_0/n_s and N are held fixed during the differentiation.

Equation (16) becomes useful when written for renormalized quantities as follows: The renormalized H is defined as in Section IV by

$$H = H_B(L^{-\varepsilon/2}Z_{H\theta}u_H, N) \ , \tag{17}$$

where the constants from (7) and (9) necessary to convert from the variable ζ_0/n_s to u_H^0 are merely absorbed into the definition of H_B. Now, substitution of (17) into (16) gives the renormalization group equation [8,12]

$$[L(\partial/\partial L) + \beta_H(u_H)(\partial/\partial u_H)]H(u_H,N,L) = 0 \ , \tag{18}$$

where the hydrodynamic Gell-Mann-Low function is defined by

$$\beta_H \equiv L(\partial u_H/\partial/L)|_F \ . \tag{19}$$

Using (9) and (10) enables (19) to be written more conveniently as

$$\beta_H/(\varepsilon/2)(u_H = (\partial \ln Z_{H\theta}/\partial u_H) \ .$$

The nonpreaveraged and preaveraged approaches, respectively, yield [12]

$$\beta_H/(\varepsilon/2)u_H = 1 - 2u_H/\varepsilon + 7(u_H)^2/3\varepsilon + O[(u_H)^3], \tag{21a}$$

$$\beta_{Hp}/(\varepsilon/2)u_{Hp} = 1 - 2u_{Hp}/\varepsilon + O[(u_{Hp})^3] \ , \tag{21b}$$

It is likely that (21b) holds to all orders in perturbation theory.

VI. Renormalization Group Scaling Analysis

Given equation (21a,b) it is straightforward to solve the linear partial differential equation (18) which only specifies the functional dependence of H on its arguments but not the function itself. This solution is found for random walk polymers to be

$$H = H(L\exp[-\int^{u_H} du_H/\beta_H(u_H)],N) \tag{22}$$

and describes the analytic dependence of large length scale dynamic properties for u_H and ε no longer small. It is readily verified by direct substitution that (22) is the general solution to (18). The solution (22) clearly involves a singular first argument of H when u_H is near one of the "fixed point" values u_H^* for which $\beta(u_H)$ vanishes. Using (21) and the condition that $u_H^* \sim O(\varepsilon)$ we find [12]

$$u_{Hp}^* = \varepsilon/2 + O(\varepsilon^3) \tag{23a}$$

$$u_H^* = (\varepsilon/2)(1 + 7\varepsilon/12) + O(\varepsilon^3) \ . \tag{23b}$$

The renormalized perturbation expansions (12) and (13) are extended to larger u_H and ε by making (12) and (13) consistent with the solution (22) and the results of simple dimensional analysis: Because L and N have the same dimensions of length and because H and u_H are dimensionless, scaling distances by $s > 0$ and then choosing $s = N$ leads to the form

$$H = H(u_H, L/N) . \tag{24}$$

Combining (22) and (24) shows H to be a function of the single hydrodynamic scaling variable

$$\xi_\theta = (2\pi N/L_H)^{\varepsilon/2} \exp[(\varepsilon/2) \int_{u_H^*/2}^{u_H} dx/\beta(x)] , \tag{25}$$

to all orders in ε. Comparing (22) and (17) and using the definition (9) and the condition that $Z_H \to 1$ as $u_H \to 0$ leads to the identity, [17]

$$\xi_\theta \equiv z_H^0/u_H^* , \tag{26}$$

of the scaling variable ξ_θ in terms of the model variable z_H^0. To first and second orders in ε we then find that (25) yields

$$\xi_\theta = (2\pi N/L)^{\varepsilon/2} \bar{u}_H (1-\bar{u}_H)^{-1} + O(\varepsilon^2) \tag{27a}$$

$$\xi_\theta = (2\pi N/L)^{\varepsilon/2} \bar{u}_H (1-\bar{u}_H)^{-1+7\varepsilon/12} + O(\varepsilon^3) \tag{27b}$$

$$\xi_\theta = (2\pi N/L)^{\varepsilon/2} \bar{u}_H (1-\bar{u}_H)^{-1} + O(\varepsilon^3) , \tag{27c}$$

where we use the definition (etc., for $\bar{u}_{Hp}$)

$$\bar{u}_H = u_H/u_H^* . \tag{27d}$$

The exact representation of (22) and (24),

$$H = H(\xi_\theta) , \tag{28}$$

combined with the approximations (27a-c) determines the general analytic dependence of (dimensionless) dynamical polymer quantities (like f/F_0) on u_H, and ε-expansions in (12) and (13a) are made consistent with (28) as follows: First we invert (27c) and (27b) to obtain

$$u_H = u_H^* \hat{\xi}_\theta (1+\hat{\xi}_\theta)^{-1} (2\pi N/L)^{-(\varepsilon/2)/(1+\hat{\xi}_\theta)} + O(\varepsilon^3) , \tag{29}$$

where the preaveraging and non-preaveraging treatments yield

$$\hat{\xi}_{\theta p} = \xi_{\theta p} \tag{30a}$$

$$\hat{\xi}_\theta = \xi_\theta (1 + \xi_\theta)^{-7\varepsilon/12} \tag{30b}$$

and $u_H(u_H^*)$ is replaced by $u_{Hp}(u_{Hp}^*)$ for the preaveraged case. Substituting (29) into (12) and (13b) gives

$$f = F_0 u_H^* \hat{\xi}_\theta (1+\hat{\xi}_\theta)^{-1} + O(\varepsilon^3) , \tag{31}$$

where again $u_H^* \to u_{Hp}^*$ in the preaveraged case. Equation (13) is derived using the condition that f scale as a power of N for N large and for $u_H = u_H^*$, a condition which readily follows from (18) and (24) for $\beta_H(u_H^*) = 0$, and which thereby motivates the reexponentiation [9]

$$1 - u_H \ln(2\pi N/L) = (2\pi N/L)^{-u_H} + O(\varepsilon^2) \tag{32}$$

for all u_H. The final result (31) is formally extended to the case where u_H and ε are not small.

It is interesting to note that forms analytically similar to (31) have previously been obtained from approximate d=3 treatments. Setting d=3 and using the definition (26) makes (31) coincide with Kirkwood's famous approximation [see equation (31.45) of Yamakawa [4].] A crossover function, which is numerically equivalent [4], is that of Debye and Bueche [21] calculated using a porous sphere model and also earlier by Hermans. [22] It is readily shown that (31) is the appropriate analytical continuation to d-space of the Kirkwood approximation and that the order ε Rouse-Zimm model agrees [23] with the order ε approximation to (31). Differences between models begin to appear in order ε^2 as is evident from (30).

Polymer chains in theta solvents, chains which are modelled by random walks, are generally found to conform to the nondraining limit of $\xi_\theta \to \infty$, a situation which emerges from (27) in the mathematical limits of <u>either</u> $u_H \to u_H^*$ or $(N/L)^{\epsilon/2} \to \infty$. The porous sphere model of Debye and Bueche [21] identifies a parameter similar to L as a length describing the penetration depth of fluid into the polymer. Hence, we obtain the nondraining limit for long random walk poymers with $N/L \to \infty$ when L ~ $O(\ell)$, corresponding to minimal penetration. Consequently, in theta solutions experiments only a limited penetration of fluid into the polymer due to its relative compactness. The situation becomes much richer when excluded volume interactions are present. The "penetration depth" L may be much larger and lead to important deviations from the nondraining limit. [17] The free-draining limit has $L \sim O(N)$ with each monomer independently disturbing the fluid flow.

The discussion here is limited to random walk polymers in order to illustrate the general concepts and methods of the dynamical renormalization group in treating polymers. When excluded volume interactions are introduced, it becomes straightforward, in principle, to evaluate the appropriate generalizations of (12) and (13b), but the solution to the corresponding renormalization group equation to order ϵ^2 becomes highly complicated. [17]

Further insight into the analytical operation to the renormalization group theory is obtained by expanding (31) in powers of ξ_θ and by using (26). Considering the preaveraging approximation, for simplicity, gives

$$f = F_0 z_H^0 \sum_{k=0}^{\infty} (-z_H/u_{Hp}^*)^k \tag{33}$$

which exactly coincides with (13a) through $(z_H^0)^3$ terms. The solution (31) for the preaveraging case is simply an Euler transform [24] of the perturbation series (13a), a method for resumming series. (The non-preaveraging case is somewhat more complicated.) This simple example nicely illustrates how the renormalization group method systematically provides a resummation of the perturbation expansions.

Equation (31) describes a standard type of crossover between the free-draining $\hat{\xi}_\theta \to 0$ and nondraining $\hat{\xi}_\theta \to \infty$ limits for random walk (or theta) polymers, but this result contrasts sharply with the renormalization group calculations of Oono which consider both hydrodynamic and excluded volume interactions and which has been claimed to produce only the nondraining limit when the excluded volume interactions are taken to vanish. [8] Oono's double crossover dependence on both of these interactions, however, becomes singular for theta chains, and the proper treatment of the limit yields (31). [17]

VII. Corrections to Preaveraging in Nondraining Limit

We now consider the friction coefficient in the nondraining limit $\hat{\xi}_\theta \to \infty$ to evaluate the leading corrections to the preaveraging approximation. Upon use of the d-space form of Stokes' law [3,8,23], $f/n_s \propto R_H^{d-2}$, the $\hat{\xi}_\theta \to \infty$ limit of (31) produces

$$R_H/R_{Hp} = (u_H^*/u_{Hp}^*)^{1/(2-\varepsilon)} = (u_H^*/u_{Hp}^*)^{1/2} + O(\varepsilon^2) \ . \qquad (34)$$

Equation (34) gives the value $R_H/R_{Hp} = 1.26 + O(\varepsilon^2)$ for $\varepsilon = 1$ (d =3). This result is roughly double the preaveraging correction of 12% found by Zimm [5] and by de la Torre et al. [6] in Monte Carlo simulations of the dynamics of rigid random walk polymers. The discrepancy must be due to the neglect of higher terms in ε. Although the renormalization group treatment generally only yields asymptotic expansions, previous experience with the calculation of static properties indicates excellent agreement [9] with both Monte Carlo simulations where the optimal order of truncation [25] for the excluded volume problem appears to be order ε^2. However, there is no guarantee that the same situation prevails with the Kirkwood-Riseman dynamical model; more work (order ε^3 or ε^4 calculations) must be done to understand this interesting point further.

VIII. Discussion

The Kirkwood-Riseman model of the dynamics of rigid polymers provides a physically simple, but mathematically complicated example of problems of the hydrodynamics of strongly interacting particles. The renormalization group is shown to introduce the essential simplification of permitting the long range hydrodynamic interactions between particles to be treated as a perturbation. The renormalization group machinery then provides the analytic continuation to strongly interacting systems using a perturbation expansion which is based on well established methods. While the leading order computation of the preaveraging correction for the Kirkwood-Riseman model hydrodynamic radius is only qualitatively correct, it does serve to illustrate the renormalization group methods in a rather simple context.

It is also relevant to mention deviations which exist between Kirkwood-Riseman model predictions and experiemental data. For instance, in theta solvents Kirkwood-Riseman theory for R_H is in error by 15 and 25% for polystyrene and poly(methylmethacrylate), respectively. [26] However, the deviations may arise from the presence of residual or ternary interactions [27] at the theta point, from the rigid body approximation of the Kirkwood-Riseman model, as well as possibly from the neglect of the dynamical solvent velocity fluctuations [8]. Hence, further work is necessary to assess the role of these three approximations, and perhaps others, on differences between the Kirkwood-Riseman model and experiment. A recent renormalization group treatment [28] of the Rouse-Zimm model of polymer dynamics shows that the order ε^2 approximation to R_H is identical with and without preaveraging and coincides with the preaveraged Kirkwood-Riseman model approximation to that order.

Acknowledgement

This research is supported, in part, by NSF Grant DMR86-14358. JFD acknowledges a fellowship in polymer science from the IBM Corporation.

Notes and References

1. P.E. Rouse, Jr., J. Chem. Phys. 21, 1272 (1953); B.H. Zimm, J. Chem. Phys. 24, 269 (1956).

2. K.F. Freed and M. Muthukumar, J. Chem. Phys. 76, 6186 (1982); M. Muthukumar and K.F. Freed, J. Chem. Phys. 76, 6195 (1982); 78, 497, 511 (1983).

3. J.G. Kirkwood and J. Riseman, J. Chem. Phys. 16, 565 (1946); J.G. Kirkwood, Macromolecules, ed. I. Oppenhiem (Gordon and Breach, New York, 1967).

4. For original references and a general discussion see H. Yamakawa, Modern Theory of Polymer Solutions (Harper and Row, New York, 1971).

5 B.H. Zimm, Macromolecules 13, 592 (1980).

6. J.G. de la Torre, A. Jimenez and J. Friere, Macromolecules 15, 148 (1982).

7. Y. Oono and K.F. Freed, J. Phys. A15, 1931 (1982).

8. Y. Oono and M.J. Kohmoto, Phys. Rev. Lett. 49, 1397 (1982); J. Chem. Phys. 78, 520 (1983); Y. Oono, J. Chem. Phys. 79, 4620 (1983). These calculations contain some order ε^2 terms but not those from the order ε^2 portion of the fixed point [12]; Y. Oono, Adv. Chem. Phys. 61, 301 (1985).

9. J.F. Douglas and K.F. Freed, Macromolecules 17, 1854, 2344 (1984); 18, 201 (1985); K.F. Freed, Accts. Chem. Res. 18, 38 (1985); K.F. Freed, Renormalization Group Theory of Macromolecules (Wiley-Interscience, New York, 1987).

10. Y. Oono and K.F. Freed, J. Chem. Phys. 75, 1009 (1981).

11. Y. Shiwa and K. Kawasaki, J. Phys. C. 15, 5349 (1982).

12. S.Q. Wang, J.F. Douglas and K.F. Freed, J. Chem. Phys. 85, 3674 (1986).

13. A more mathematical introduction to the conformations of polymers is given in K.F. Free, Ann. Prob. 9, 537 (1981).

14. H. Fikentscher and H. Mark, Kolloid Z., 49, 185 (1929); W. Kuhn, Kolloid Z, 68, 2 (1934); W. Kuhn and H. Kuhn, Helv. Chim. Acta, 26, 1394 (1943); T. Fox and P.J. Flory, J. Am. Chem. Soc. 73, 1904 (1951).

15. J.F. Douglas and K.F. Freed, Macromolecules 17, 2354 (1984).

16. S.Q. Wang, J.F. Douglas and K.F. Freed, Macromolecules 18, 2464 (1985).

17. S.Q. Wang, J.F. Douglas and K.F. Freed, "Influence of Variable Draining and Excluded Volume on the Hydrodynamic Radius within the Kirkwood-Riseman Model" (submitted).

18. Alternatively, expression for $\underset{\approx}{T}$ may be introduced in which corrections are made for the nonzero size of the "monomers". See J. Rotne and S.J. Prager, J. Chem. Phys. 50, 4831 (1969). The d-space form of these corrections can, however, be shown to make negligible contributions when $N \to \infty$. They are irrelevant variables in renormalization group terminology.

19. Care must be exercised in ignoring the cut-off for shorter chains as described by K. Osaki, Macromolecules 5, 141 (1972).

20. P. Ramond, Field Theory, A Modern Primer (Benjamin/Cummings, Reading, Mass., 1981).

21. P. Debye and A.M. Bueche, J. Chem. Phys. 15, 573 (1948); P. Debye, Phys. Rev. 71, 487 (1947).

22. J.J. Hermans Rec. Trav. Chim. 63, 219 (1944).

23. S. Stepanow, J. Phys. A17, 3041 (1984); G.F. Al-Noimi, G.C. Martinez-Mekler and C.A. Wilson, J. Phys. 39, L373 (1978).

24. G.H. Hardy, Divergent Series (Clarendon Press, Oxford, 1949) pp. 197-8.

25. J.F. Douglas, S.Q. Wang and K.F. Freed, Macromolecules 19, 2207 (1986).

26. W. Burchard and M. Schmidt, Macromolecules 14, 210 (1981); H.U.ter Meer, W. Burchard and W. Wunderlich, Coll.Polym. Sci. 258, 675 (1980).

27. See reference cited in 19.

28. S.Q. Wang and K.F. Freed, J. Chem, Phys. 85, 6210 (1986).

ON THE HYDRODYNAMIC LIMIT OF A SCALAR GINZBURG-LANDAU LATTICE MODEL: THE RESOLVENT APPROACH

J. Fritz
Mathematical Institute HAS
H-1364 Budapest, POB 127, Hungary

Abstract

A d-dimensional lattice system of continuous spins is considered, the evolution law is given by an infinite system of locally coupled stochastic differential equations. In view of the construction of the model, the evolution is reversible in time, and the mean spin satisfies a conservation law, thus we have a whole family of equilibrium Gibbs states parametrized by the associated chemical potential. The asymptotic behaviour of the system is investigated in the framework of hydrodynamic (Navier-Stokes) scaling. The initial, local equilibrium distribution is specified by letting the chemical potential of the equilibrium state be inhomogeneous in space. We show that the macroscopic fluctuations vanish in the hydrodynamic limit, and the limiting profile of the mean spin satisfies a nonlinear diffusion equation. Some very recent results and the main ideas of the proofs are summarized.

0. Introduction

The idea that the equations of fluid mechanics can be derived from microscopic laws is certainly not new, for a fairly convincing intuitive derivation of the Euler equations from classical statistical mechanics see MORREY [24]. According to the general philosophy behind the argument, these familiar macroscopic laws can be extracted from the enormously complex world of the dynamics of microscopic particles by means of a procedure usually referred to as the hydrodynamic scaling limit. In this setup the law of large numbers implies the convergence of the rescaled extensive quantities of the microscopic system

---

Supported by Universität Heidelberg, SFB 123 and by the University of Minnesota, Institute for Mathematics and its Applications (IMA)

to their local equilibrium expectations as the scaling parameter, the ratio of the microscopic and macroscopic units of time and space, goes to zero. Since the equilibrium distribution is specified by the mean values of the conserved quantities, exactly these quantities are expected to satisfy a closed system of evolution equations.

A mathematical formulation of the principles of hydrodynamic scaling in terms of infinite systems and Gibbs random fields was presented in 1978 by Dobrushin [9], but the text of this talk has never been published, cf. [3,4,10]. Some relevant ingredients of the mathematical theory of hydrodynamic scaling appeared in the papers by ROST [28] and by KIPNIS-MARCHIORO-PRESUTTI [22], for a survey of ideas and results see the lecture notes by DE MASI-IANIRO-PELLEGRINOTTI-PRESUTTI [6]. These results are based on a very good understanding of the underlying microscopic model allowing more or less explicit calculations to prove the principle of local equilibrium. The main purpose of this paper is to develop methods for models containing a functional parameter. This level of generality usually excludes the possibility of explicit calculations; we are not able to prove the principle of local equilibrium directly. Of course, the models we can handle still do have a very particular structure I shall try to expose later. To show that we really can pass to the hydrodynamic limit, the resolvent equation approach of PAPANICOLAU-VARADHAN [25] will be adapted to the present situation. This extremely flexible technics applies, in principle to various problems concerning different sorts of reversible processes, compare e.g. [20] and [26]. The information we need on the microscopic dynamics is mainly qualitative, we have to understand that the dependence of the evolution on initial data is smooth in a well defined sense. The related a priori bounds can be proven by exploiting the nice parabolic structure of the microscopic dynamics. The first information on this structure is certainly the form of the limiting equation, namely

$$\frac{\partial \rho(t,x)}{\partial t} = \frac{1}{2} \operatorname{div} [D(\rho) \operatorname{grad} \rho(t,x)] \quad , \tag{0.1}$$

where D is strictly positive and bounded.

The Ginzburg-Landau model we are going to discuss is a phenomenological but still microscopic model of magnetization, see [21,23,32] with some further references . Since the precession of the spins in their magnetic field will be neglected, we may and do assume that we are given a scalar field. Consider now a system $S_k \varepsilon R$, $k \varepsilon Z^d$ of block spins of a microscopic system; e.g. let S_k be defined as the integral mean of the microscopic field on the unit cube, W_k with center $k \varepsilon Z^d$. The whole configuration of spins will be denoted by $S = (S_k)_{k \varepsilon Z^d}$. The cells W_k are assumed to be so large that we can speak about a thermodynamical system in each of them, but they are so small that the whole system is in a state of local (cellular) equilibrium, thus the thermodynamical potentials are local functions of the configuration. Suppose that the system is in a thermal equilibrium, and the free energy of cell W_k equals

$$H_k(S) = V(S_k) + \frac{\alpha}{2} \sum_{J:|j-k|=1} (S_k - S_j)^2 \quad , \tag{0.2}$$

where $\alpha > 0$ and $V : R \to R$ is convex; then the formal expression for the free energy of the whole system is

$$H(S) = \sum_{k \varepsilon Z^d} [V(S_k) + \frac{\alpha}{4} \sum_{|j-k|=1} (S_k - S_j)^2] \ . \tag{0.3}$$

The chemical potential, m_k associated with the magnetization S_k in W_k is given by

$$m_k = \frac{\partial H_k}{\partial S_k} = \frac{\partial H}{\partial S_k} = V'(a_k) - \alpha \sum_{|j-k|=1} (S_j - S_k) \quad , \tag{0.4}$$

where V' denotes the derivative of V . Since our system is not homogeneous in space, there is a current of magnetization between neighboring cubes. In a thermal equilibrium the intensity of this current is proportional to the gradient of the chemical potential, while the microscopic effects that have been neglected so far can be described by random, completely uncorrelated currents. More exactly, the stochastic differential of the current of magnetization from W_k to W_j , $|j-k|=1$, is given by (0.5)

$$dJ_{kj} = \frac{1}{2} (m_k - m_j)\, dt - dw_{kj} ,$$

where w_{kj} is a family of independent standard Wiener processes defined for the positively oriented bonds $k \to j$ of Z^d , and $w_{kj} = -w_{jk}$ if $k \to j$ is a negatively oriented bond. Then the intensity of the current flowing out of W_k is just

$$\operatorname{div} dJ_k = \sum_{j:|j-k|=1} dJ_{kj} ,$$

consequently the evolution law reads as

$$dS_k + \operatorname{div} dJ_k = 0 . \tag{0.7}$$

In a more explicit form we have

$$dS_k = \frac{1}{2} \sum_{|j-k|=1} \left(\frac{\partial H}{\partial S_j} - \frac{\partial H}{\partial S_k}\right) dt + \sum_{|j-k|=1} dw_{kj} \tag{0.8}$$

with initial condition $S_k(0) = \sigma_k$, $k \in Z^d$; while

$$dS = \frac{1}{2} \operatorname{div} \operatorname{grad} \frac{\partial H}{\partial S}\, dt + \operatorname{div} dw = \frac{1}{2} \Delta \frac{\partial H}{\partial S}\, dt + \nabla^* dw \tag{0.9}$$

are more expressive, short hand formulations of the same law.

From a mathematical point of view, (0.8) is the simplest reversible system satisfying a conservation law. Indeed, the formal generator of a reversible process can be represented as

$$Gf = \frac{1}{2} \sum_k \sum_j \exp(H_k(\sigma)) \frac{\partial}{\partial \sigma_k} \left[\exp(-H_k(\sigma))\, a_{kj}(\sigma) \frac{\partial f(\sigma)}{\partial \sigma_j}\right] , \tag{0.10}$$

where $a = (a_{kj})$ is a positive definite matrix for each configuration,

$\sigma = (\sigma_k)_{k \varepsilon Z^d}$; the generator of the process defined by (0.8) corresponds to the diffusion matrix $a = -\Delta$, where $\Delta = (\Delta_{kj})$ denotes the matrix of the discrete Laplace operator, cf. (0.9); i.e. $\Delta_{kk} = -2d$, $\Delta_{kj} = 1$ if $|j-k|=1$, $\Delta_{kj} = 0$ otherwise, thus

$$\Delta\phi_k = \sum_{j \varepsilon Z^d} \Delta_{kj}\phi_j = \sum_{j:|j-k|=1} (\phi_j - \phi_k) \tag{0.11}$$

whenever $\phi = (\phi_k)_{k \varepsilon Z^d}$. Moreover, if μ denotes the Gibbs state with interaction H at unit temperature, then the stationary Kolmogorov equation $\int Gf\, d\mu = 0$ happens to be a formal consequence of (0.10), thus μ is a stationary measure whenever the dynamics is well defined. The converse problem is much more complex, see the related results of ANDJEL [2] for zero range processes. Nevertheless, calculating the time-derivative of the free energy of a stationary measure, we see that there is only one stationary measure if the diffusion matrix $a = (a_{kj})$ is strictly positive definite, but there are many other stationary states if 0 is an eigenvalue of $a = (a_{kj})$.

More exactly, let μ_λ denote the Gibbs state for the interaction $H_\lambda(\sigma) = H(\sigma) - \sum \lambda_k \sigma_k$ whenever $\lambda = (\lambda_k)_{k \varepsilon Z^d}$ is a real sequence indexed by Z^d , then the conditional distribution of σ_k given the rest $\sigma_k^c = (\sigma_j)_{j \neq k}$ of the configurations has the Gibbs form

$$\mu(d\sigma_k | \sigma_k^c) = \frac{1}{Z(\lambda_k, \sigma_k^c)} \exp[-H_k(\sigma) + \lambda_k \sigma_k]\, d\sigma_k \ , \tag{0.12}$$

for each k , where

$$Z(u, \sigma_k^c) = \int_{-\infty}^{+\infty} \exp[-H_k(\sigma) + u\sigma_k]\, d\sigma_k \ . \tag{0.13}$$

Integrating Gf by parts we obtain the fundamental identity

$$\int Gf\, d\mu_\lambda = -\frac{1}{2} \sum_k \int (\sum_j a_{kj}(\sigma)\lambda_j) \frac{\partial f(\sigma)}{\partial \sigma_k}\, \mu_\lambda(d\sigma) \quad , \tag{0.14}$$

consequently μ_λ is a stationary measure if

$$\sum_{j \in Z^d} a_{kj}(\sigma)\lambda_j = 0 \quad \text{for all } \sigma \text{ and } k ; \tag{0.15}$$

the converse problem is open.

The interpretation of the parameter $\lambda = (\lambda_k)$ is immediate; integrating (0.13) by parts we obtain

$$\lambda_k = \int \frac{\partial H(\sigma)}{\partial \sigma_k} \mu_\lambda(d\sigma) , \tag{0.16}$$

thus λ_k is the mean value of the quasi-microscopic chemical potential m_k defined by (0.4). Let $\rho = (\rho_k)_{k \in Z^d}$ denote the mean spin with respect to μ_λ, then we have

$$\rho_k = \int \rho_k \, \mu_\lambda(do) = \frac{\partial}{\partial u} F_k(u,\lambda) \quad \text{at } u = \lambda_k , \tag{0.17}$$

where the local free energy

$$F_k(u,\lambda) = \int \log Z(u,\sigma_k^c) \, \mu_\lambda(d\sigma) \tag{0.18}$$

is strictly convex function of the chemical potential $u = \lambda_k$, thus we have a one-to-one correspondence between λ and ρ.

Remember that the physical picture contains an underlying microscopic system with a completely unspecified structure. From now on the system of block spins $S = (S_k)_{k \in Z^d}$ and its evolution (0.8) will be referred to as the microscopic system; the initial configuration $S(0) = \sigma$ will be distributed by a local equilibrium distribution μ_λ, $\lambda = (\lambda_k)_{k \in Z^d}$. In view of (0.17) the extensive field associated to the chemical potential λ is just the mean spin ρ. On the other hand, the compact form (0.7) of the evolution equation is in fact the conservation law for the spin S, thus we expect that passing to the hydrodynamic limit a closed equation will be obtained for the mean spin ρ. Indeed, the principle of local equilibrium means that the basic thermodynamics (0.16) - (0.18) remains in force with a time-dependent profile λ of the chemical potential, thus

(0.16) yields $\frac{\partial \rho}{\partial t} = \frac{1}{2} \Delta\lambda = \frac{1}{2} \operatorname{div} (\frac{\partial \lambda}{\partial \rho} \operatorname{grad} \rho)$ for the limiting value of the mean spin. The functional dependence between ρ and λ is given by the limiting form $\rho = F'(\lambda)$ of (0.17), where $F(u)$ is the free energy for a homogeneous profile $\lambda_k = u$. Therefore, we have (0.1) with $D(\rho) = \frac{\partial \lambda}{\partial \rho} = 1/F''(\lambda)$ if $\rho = F'(\lambda)$. This coefficient of diffusion is the very same as the variance of equilibrium fluctuations, see SPOHN [32].

1. The Hyrodynamic and the Continuum Scaling Rules

The goal of the hydrodynamic scaling is to give a mathematical interpretation to such obvious, experimental facts that the number of particles in macroscopically small domains is very large, and the relaxation to equilibrium of such subsystems is extremely fast. More exactly, we have two different scales of time and space: If $t \geq 0$ and $x \in R^d$ denote the macroscopic time and position, then the microscopic time goes as $\tau = t/\varepsilon^2$, while the microscopic position turns into $k = [x/\varepsilon]$, where the scaling parameter $\varepsilon > 0$ goes to zero, $[x/\varepsilon]$ denotes the vector $K \in Z^d$ formed by the integer parts of the components of x/ε. The anomalous scaling of the time is due to the diffusive nature of the evolution (0.8), see [6]. According to this scaling rule, we are interested in the asymptotic behaviour of the rescaled density field

$$S_\varepsilon(t,\phi) = \int \phi(x)\, S_\varepsilon(t,x)\, dx \quad , \tag{1.1}$$

where ϕ is a test function,

$$S_\varepsilon(t,x) = S_{[x/\varepsilon]}(t/\varepsilon^2) \quad , \tag{1.2}$$

and the initial configuration of the evolution (0.8) is distributed by a local equilibrium distribution $\mu_{\lambda,\varepsilon}$ of type (0.12) with profile $\lambda_\varepsilon = (\lambda_{\varepsilon,k})_{k \in Z^d}$ defined by

$$\lambda_{\varepsilon,k} = \varepsilon^{-d} \int_{[x/\varepsilon]=k} \lambda(x)\, dx \quad . \tag{1.3}$$

This means that the initial condition is specified in terms of a continuous profile $\lambda : R^d \to R$.

To make the heuristic argument at the end of the Introduction precise, we have to show that

$$\rho_\varepsilon(t,\phi) = E_{\lambda,\varepsilon}(S_\varepsilon(t,\phi)) \underset{\varepsilon \to 0}{\to} \int \phi(x)\rho(t,x)\, dx \tag{1.4}$$

and

$$E_{\lambda,\varepsilon} \int (\Delta_\varepsilon\phi(x))[V'(S_\varepsilon(t,x)) - \alpha\varepsilon^2\, \Delta_\varepsilon S_\varepsilon(t,x)]\, dx \tag{1.5}$$

$$\underset{\varepsilon \to 0}{\to} \int (\Delta\phi(x))\ \lambda(t,x)\, dx$$

with some measurable functions ρ and λ related by

$$\rho(t,x) = F'(\lambda(t,x)) \quad . \tag{1.6}$$

Here $E_{\lambda,\varepsilon}$ denotes the expectation with respect to the stochastic dynamics (0.8) with initial distribution $\mu_{\lambda,\varepsilon}$,

$$F(u) = \lim_{\Lambda \to R^d} \frac{1}{\Lambda} \log \int \exp[-H_\Lambda(\sigma) + u \sum_{k\in\Lambda} \sigma_k] \prod_{k\in\Lambda} d\sigma_k \tag{1.7}$$

denotes the free energy of the translation invariant Gibbs state u_u^o with Hamilton function $H_u(\sigma) = H(\sigma) - u \sum \sigma_k$, $u \in R$, while Δ is the usual Laplace operator and

$$\Delta_\varepsilon f(x) = \varepsilon^{-2} \sum_{i=1}^{d} (f(x+\varepsilon e_i) + f(x-\varepsilon e_i) - 2f(x)) \tag{1.8}$$

whenever $f : R^d \to R$; $e_1,e_2,\ldots,e_d$ are the unit vectors of our orthogonal system of coordinates in R^d . Notice that the second factor of the integrand on the left hand side of (1.5) is just the gradient $\partial H/\partial S_k$ of the hamiltonian in the given scaling. The macroscopic version (1.4) - (1.7) of the principle of local equilibrium is somewhat weaker than the usual one, see [6] and the discussion below.

To continue the discussion of the hydrodynamic limit and its relation to the continuum (lattice approximation) scaling we need the following technical framework. Let L denote the space of locally integrable $f : R^d \to R$, the space of step-functions of step-size ε will be denoted by L_ε ; $f \varepsilon L_\varepsilon$ means that $f(x) = f(y)$ if $[x/\varepsilon] = [y/\varepsilon]$. At a given level $\varepsilon > 0$ of scaling the sequences $\sigma = (\sigma_k)_{k \varepsilon Z^d}$ will be embedded into L_ε by the one-to-one correspondence $\sigma_\varepsilon(x) = \sigma_{[x/\varepsilon]}$. Let $W_\varepsilon(x)$ denote the cell of εZ^d containing x , i.e. $W_\varepsilon(x) = [y:[y/\varepsilon] = [x/\varepsilon]]$, then the operator I_ε of integral mean $\lambda_\varepsilon = I_\varepsilon \lambda$ given by

$$\lambda_\varepsilon(x) = \varepsilon^{-d} \int_{W_\varepsilon(x)} \lambda(y)\, dy \tag{1.9}$$

is a projection of L onto L_ε .

The configuration space in which our stochastic dynamics lives in the most natural way can be defined by means of the following sequence, $|\cdot|_r$, $r \varepsilon N$ of Hilbert norms

$$|\lambda|_r^2 = \int \exp(- \frac{|x|}{r}) \, \lambda^2(x) dx \; ; \tag{1.10}$$

then $H_r \subset L$ is the Hilbert space with norm $|\cdot|_r$, and $H_e = \bigcap H_r$. In the strong topology of H_e the relation $\lim \lambda_n = 0$ means that $\lim |\lambda_n|_r = 0$ for each $r \varepsilon N$. The dual space, H_e^* of H_e with respect to $L_2(R^d)$ is defined by means of the norms $|\cdot|_{-r}$,

$$|\phi|_{-r}^2 = \int \exp(\frac{|x|}{r}) \, \phi^2(x) \, dx \, . \tag{1.11}$$

Indeed, if H_{-r} , $r \varepsilon N$ is the Hilbert space with norm $|\cdot|_{-r}$, then $H_e^* = UH_{-r}$; and $\lim \phi_n = 0$ in H_e^* means that $\lim |\phi_n|_{-r} = 0$ for some $r \varepsilon N$. If $\phi \varepsilon H_e^*$ then $\phi(\lambda) = \int \phi(x)\lambda(x)\, dx$ for all $\lambda \varepsilon H_e$. The space of twice weakly differentiable $\lambda \varepsilon H_e$ such that all first and second derivatives of λ also belong to H_e will be denoted by H_e^2 . A nonlinear mapping $f : H_e \to R$ is continuously differentiable if there exists a mapping $Df : H_e \to H_e^*$ such that $Df(\sigma,\delta) = \int \delta(x)\, Df(\sigma,x)\, dx$ is a continuous function of $\sigma \varepsilon H_e$ for each $\delta \varepsilon H_e$,

and

$$f(\sigma+\delta) = f(\sigma) + {}_0\int^1 \int \delta(x)\, Df(\sigma+t\delta)\, dx\, dt \tag{1.12}$$

for all $\sigma,\delta\varepsilon H_e$. The functional $Df(\sigma,\cdot)$ is uniquely determined by (1.12) for each σ , it will be called the functional derivative of f at $\sigma\varepsilon H_e$.

Besides the scalar fields $\lambda,\sigma,S_\varepsilon \,\varepsilon\, L$ some vector fields $\underline{\lambda} : R^d \to R^d$ play also some role. The space of vector fields with locally integrable components will be denoted by L^d . If $\lambda\varepsilon L$ then its gradient of step-size ε is in fact a vector field defined by

$$\nabla_\varepsilon \lambda(x) = \frac{1}{\varepsilon} \sum_{i=1}^{d} [\lambda(x+\varepsilon e_i) - \lambda(x)]\, e_i \quad . \tag{1.13}$$

On the other hand, if $\underline{\lambda}\varepsilon L^d$ and $<\cdot,\cdot>$ denotes the scalar product in R^d then

$$\nabla_\varepsilon^* \underline{\lambda}(x) \;(= \frac{1}{\varepsilon} \sum_{i=1}^{d} < e_i, \underline{\lambda}(x) - \underline{\lambda}(x-\varepsilon e_i) > \tag{1.14}$$

defines the divergence of $\underline{\lambda}$ of step-size ε . If $u\varepsilon L$ and $\underline{v}\varepsilon L^d$ vanish at infinity, then

$$\int < \underline{v}(x), \nabla_\varepsilon u(x) > dx = - \int u(x)\, \nabla_{\underline{\varepsilon}}^* v(x)\, dx \quad , \tag{1.15}$$

i.e. ∇_ε^* is the formal adjoint of ∇_ε , and $\nabla_\varepsilon = \nabla_\varepsilon^* \nabla_\varepsilon$.

The current (0.5) is a typical vector field, its components are the currents flowing out along the positively oriented bonds. Using the above notation, the evolution law (0.8) should be rewritten for the rescaled spin system $S_\varepsilon = S_\varepsilon(t,x) = S_{[x/\varepsilon]}(t/\varepsilon^2)$ as follows. Let $\underline{w}_\varepsilon = \underline{w}_\varepsilon(t,x)$ denote a stochastic processes such that $\underline{w}_\varepsilon(t,x) = \underline{w}_\varepsilon(t,y)$ if $[x/\varepsilon] = [y/\varepsilon]$, and $w_k(t) = \underline{w}_\varepsilon(t,\varepsilon k)$, $k\varepsilon Z^d$ is a family of independent, standard Wiener processes in R^d , then

$$dS_\varepsilon = \frac{1}{2} [\Delta_\varepsilon V'(S_\varepsilon) - \alpha\varepsilon^2 \Delta_\varepsilon^2 S_\varepsilon]\, dt + \nabla_\varepsilon^* \underline{w}_\varepsilon(dt,x) \tag{1.16}$$

with initial condition $S_\varepsilon(0,\cdot) = \sigma_\varepsilon \in H_e \cap L_\varepsilon$. In particular from the point of view of the continuum limit, it is convenient to specify the initial configuration as the projection $\sigma_\varepsilon = I_\varepsilon \sigma$ of a fixed $\sigma \in H_e$. Similarly, the initial distribution $\mu_{\lambda,\varepsilon}$ can be considered as a probability measure on the Borel sets of H_e ; of course $\mu_{\lambda,\varepsilon}(H_e \cap L_\varepsilon) = 1$. Finally, let P_ε^t and G_ε denote the Markov semigroup and its formal generator associated with (1.16). Any f_ε : $H_e \cap L_\varepsilon \to R$ can be extended to H_e by $f(\sigma) = f(I_\varepsilon \sigma)$; if $f : H_e \to R$ is continuously differentiable then our fundamental identity (0.14) turns into

$$\int G_\varepsilon f(\sigma)\, \mu_{\lambda,\varepsilon}(d\sigma) = \frac{1}{2} \int \int [\Delta_\varepsilon \lambda(x)]\, Df(\sigma,x)\, dx\, \mu_{\lambda,\varepsilon}(d\sigma) \tag{1.17}$$

Suppose now that

$$f(\sigma) = h(\phi_1(\sigma),\phi_2(\sigma),\ldots,\phi_n(\sigma)) \, , \tag{1.18}$$

where $h : R^n \to R$ has continuous and bounded first derivatives, $\phi_1,\phi_2,\ldots,\phi_n$ belong to H_e^* , while $\phi_k(\sigma) = \int \phi_k(x)\sigma(x)\, dx$; in this case it is easy to show that

$$\lim_{\varepsilon \to 0} \int f(\sigma)\, \mu_{\lambda,\varepsilon}(d\sigma) = f(\rho_0) \tag{1.19}$$

with $\rho_0(x) = F'(\lambda(x))$, see (1.7) , while

$$\lim_{\varepsilon \to 0} \int G_\varepsilon f(\sigma)\, \mu_{\lambda,\varepsilon}(d\sigma) = \frac{1}{2} \int Df(\rho_0,x)\, \Delta\lambda(x)\, dx = G_0 f(\rho_0) \, , \tag{1.20}$$

at least if $\lambda \in H_e^2$. Observe that the operator G_0 defined by the second quality of (1.20) is just the formal generator of the semigroup association with (0.1). This means that if σ is a "typical" configuration with respect to the family $\mu_{\lambda,\varepsilon}$ of initial distributions, then $G_\varepsilon f(\sigma) \to G_0 f(\sigma)$, therefore we are tempted to refer to a Trotter-Kurtz theorem. Of course, we are not able to give a satisfactory definition of the phrase "typical", but it is remarkable that this heuristic argument does not rely on the principle of local equilibrium for positive times. Indeed, the relations (1.19) and (1.20) will play a crucial role in the derivation of (0.1) as the hydrodynamic limit of (0.8).

The rest of this section is devoted to an explanation of the difference between the procedures of hydrodynamic and continuum limit, we consider a slightly more general problem than that of (0.8). Let $\underline{J} : R \to R^d$ be continuously differentiable with bounded first derivatives, and let $S_\varepsilon = S_\varepsilon(t,x,\sigma)$ denote the solution to

$$dS_\varepsilon = \Delta_\varepsilon^* \underline{J}(S_\varepsilon)\, dt + \frac{1}{2} \Delta_\varepsilon V'(S_\varepsilon)\, dt - \frac{\alpha}{2} \varepsilon^a \Delta_\varepsilon^2 S_\varepsilon\, dt + \beta\varepsilon^{-b} \nabla_\varepsilon^* \underline{w}_\varepsilon (dt,x) \tag{1.21}$$

with initial condition $S_\varepsilon(0,x,\sigma) = I_\varepsilon \sigma$. We are interested in the asymptotic behavior of the density field

$$S_\varepsilon(t,\phi,\sigma) = \int \phi(x)\, S_\varepsilon(t,x,\sigma)\, dx \tag{1.22}$$

as $\varepsilon \to 0$ while $\sigma \varepsilon H_e$ is fixed, a very different behavior can be expected for different values of the parameters $\alpha > 0$, $\beta \varepsilon R$, $0 \leqslant a \leqslant 2$, $0 \leqslant b \leqslant d/2$. The additional term $\Delta_\varepsilon^* \underline{j}$ corresponds to an external driving force $\underline{J}$ in (0.5). In the picture of latice approximation the interaction should also be rescaled, this means a formal hamiltonian

$$H_\varepsilon(\sigma_\varepsilon) = \int [V(\sigma_\varepsilon) + \frac{\alpha}{2} |\nabla_\varepsilon \sigma_\varepsilon|^2]\, dx \quad , \tag{1.23}$$

and $a = 0$ in (1.21). The case $b = d/2$ is distinguished by the requirement that $S_\varepsilon(t,\phi,\sigma)$ should have a proper, nonsingular stochastic differential for smooth ϕ. Indeed, if $\underline{J}$ and V' are linear, and $\beta > 0$, $b = d/2$, then S_ε converges to a nondegenerate Gauss field for each $\sigma \varepsilon H_e$, but this limiting field is a generalized one if $\alpha = 0$ or $d > 1$.

The deterministic case when $\beta = 0$ is more or less clear, too. If $d = 1$, $J = 0$, $\alpha = \beta = 0$ and V''' is bounded while

$$0 < c_1 \leqslant V''(x) \leqslant c_2 < +\infty \tag{1.24}$$

with some constants c_1 and c_2 , then we prove in [16] that S_ε converges weakly to the corresponding solution to

$$\frac{\partial \rho}{\partial t} = \frac{1}{2} \frac{\partial}{\partial x} (V''(\rho) \frac{\partial \rho}{\partial x}) \ . \tag{1.25}$$

In the general deterministic case, i.e. when $d \geq 1$, $\alpha \geq 0$, $a = 0$ and $\beta = 0$ the technics of [16] is still applicable, under (1.24) we expect that S_ε converges weakly to the solution to

$$\frac{\partial \rho}{\partial t} = \operatorname{div} \underline{J}(\rho) + \frac{1}{2} \operatorname{div} (V''(\rho) \operatorname{grad} \rho) - \frac{\alpha}{2} \Delta^2 \rho \ . \tag{1.26}$$

For this approach the positivity of α is not relevant. If $a > 0$ then the last term of (1.26) disappears as $\varepsilon \to 0$.

If $\alpha > 0$ then the smoothing effect of $-\Delta^2$ can also be exploited. Let $q_\varepsilon(t,x)$ denote the solution to $\partial q/\partial t = -\frac{\alpha}{2} \Delta_\varepsilon^2 q$ with initial condition $q_\varepsilon(0,x) = \varepsilon^{-d}$ if $[x/\varepsilon] = 0$ and $q_\varepsilon(0,x) = 0$ otherwise, then (1.21) can be rewritten as

$$\begin{aligned} S_\varepsilon(t,x) &= \int q_\varepsilon(t\varepsilon^a,y)\, \sigma(x-y)\, dy \\ &+ \int_0^t \int < \nabla_\varepsilon q_\varepsilon(s\varepsilon^a,y) \ , \ \underline{J}(S_\varepsilon(t-s,x-y)) > dy\, ds \\ &+ \frac{1}{2} \int_0^t \int [\Delta_\varepsilon q_\varepsilon(s\varepsilon^a,y]\, V'(S_\varepsilon(t-s,x-y))\, dy\, ds \\ &+ \beta\varepsilon^{-b} \int \int_0^t < \nabla_\varepsilon q_\varepsilon(t\varepsilon^a - s\varepsilon^a, x-y) \ , \ \underline{w}_\varepsilon(ds,y) > dy \ . \end{aligned} \tag{1.27}$$

Since $\underline{J}$ and V' are uniformly Lipschitz continuous, it is not difficult to obtain L_p - estimates for (1.2) . If $a = 0$ and $b = d/2$ then (1.2) has a proper continuum limit obtained by replacing ∇_ε , Δ_ε and $\varepsilon^{-d/2}\, \underline{w}_\varepsilon(ds,y)\, dy$ by grad , Δ and $\underline{w}(ds,dy)$, respectively, where $\underline{w}$ is a homogeneous white noise in space and time, thus the limiting equation makes sense only if grad q is square integrable, i.e. if $d=1$. In this case, the existence and uniqueness of solutions to the limiting equation was shown by FUNAKI [19], his method seems to be sufficient for passing to the limit in (1.27). If $\alpha > 0$, $a = 0$ but $b < d/2$ then S_ε is expected to converge to a deterministic limit, namely to the corresponding solution to (1.26). In view of the estimates of [19] this is quite plausible if $d = 1$, but $\Delta_\varepsilon q_\varepsilon$ becomes more singular if $d > 1$, thus the gap between b and $d/2$ might play a role, too.

Remember now that $b = 0$ and $a = 2$ in the case of the hydrodynamic limit, thus the stochastic integral of (1.27) vanishes as the deterministic integrals exhibit a singular behaviour, therefore, it is not easy to tell what happens to $S_\varepsilon(t,\phi,\sigma)$ if $\varepsilon \to 0$ while $\sigma \varepsilon H_e$ is kept fixed. Let $\rho(t,x,\lambda)$ denote the solution to (0.1) with initial condition $\rho(0,x,\lambda) = F'(\lambda(x))$ and consider the limiting density field $\rho(t,\phi,\lambda) = \int \phi(x)\, \rho(t,x,\lambda)\, dx$. We believe that for each $\lambda \varepsilon H_e$ there exists a set $C_\lambda \subset H_e$ that might be called the class of $\mu_{\lambda,\varepsilon}$- typical configurations, and $S_\varepsilon(t,\phi,\sigma) \to \rho(t,\phi,\lambda)$ whenever $\sigma \varepsilon C_\lambda$. Otherwise, the limit may be a random quantity, or it does not exist at all.

2. The Main Result and the Idea of its Proof

In addition to (1.24) and to the boundedness of V''' , due to some technical reasons we have to assume that V' is close to a linear function in the sense that

$$\frac{c_2-c_1}{c_1+c_2} < \bar{\alpha} , \tag{2.1}$$

where $\bar{\alpha}$ is a universal constant. Using notation of the previous section, we have the following

Theorem. Consider the rescaled density field $S_\varepsilon(t,\phi)$ of the Ginzburg-Landau model (0.8), and suppose that the initial configuration is distributed by $\mu_{\lambda,\varepsilon}$ for $\varepsilon \varepsilon (0,1]$. If $\lambda \varepsilon H_e^2$ then

$$S_\varepsilon(t,\phi) \underset{\varepsilon \to 0}{\to} \int \phi(x)\rho(t,x)\,dx \quad \text{in probability}$$

for each $t > 0$ and $\phi \in H_e^*$, where the limiting density $\rho(t,x)$ satisfies (0.1) with initial condition $\rho(0,x) = F'(\lambda(x))$. The free energy F is defined by (1.7), while $D(\rho) = 1/F''(\lambda)$ if $\rho = F'(\lambda)$. The diffusion coefficient D is an analytic function of ρ, and $0 < \bar{c}_1 < D(\rho) < \bar{c}_2 < +\infty$ for each ρ, the constants $\bar{c}_1$ and $\bar{c}_2$ depend only on c_1 and c_2 of (1.24). |||

The proof is based on the following variant of the resolvent equation approach of [20,25,26]. Let $g : H_e \to R$ be a function of type (1.18), e.g. $g(\sigma) = \phi(\sigma)$ with some $\phi \in H_e^*$, and introduce the resolvent

$$f_\varepsilon(\sigma_\varepsilon) = R_{z,\varepsilon} g(\sigma) = \int_0^\infty e^{-zt} P_\varepsilon^t g(\sigma)\,dt \tag{2.2}$$

for $z > 0$; here P_ε^t is the operator of conditional expectation given the initial configuration. Integrating the resolvent equation

$$g(\sigma) = z\, f_\varepsilon(\sigma) - G_\varepsilon f_\varepsilon(\sigma) \tag{2.3}$$

and using the basic identity (1.17), we obtain

$$\int g(\sigma)\,\mu_{\lambda,\varepsilon}(d\sigma) = z \int d_\varepsilon(\sigma)\,\mu_{\lambda,\varepsilon}(d\sigma) \tag{2.4}$$

$$- \frac{1}{2} \iint Df_\varepsilon(\sigma,x)\,\Delta_\varepsilon\lambda(x)\,dx\,\mu_{\lambda,\varepsilon}(d\sigma) .$$

In view of (1.19) and (1.20) we want to pass to the limiting resolvent equation

$$g(\rho_0) = z\, f(\rho_0) - \frac{1}{2} \int Df(\rho_0,x)\,\Delta\lambda(x)\,dx \tag{2.5}$$

along some subsequences, where $\rho_0(x) = F'(\lambda(x))$. Since (2.5) is just the resolvent equation of the semigroup defined by (0.1), and this equation has a unique solution, (2.5) implies

$$f(\rho_0) = \int_0^\infty e^{-zt}\, g(\rho(t,\cdot))\, dt \ , \tag{2.6}$$

where $\rho(t,x)$ solves (0.1) with initial condition $\rho(0,x) = \rho_0(x)$. Therefore, if the convergent subsequence we can select does not depend on λ, we have

$$\lim_{\varepsilon \to 0} \int f_\varepsilon(\sigma)\, \mu_{\lambda,\varepsilon}(d\sigma) = f(\rho_0) \ , \tag{2.7}$$

whence the Theorem follows by passing to the inverse Laplace transform.

This argument is very general, it can be applied even to the non-reversible model (1.21) as well as to hamiltonian systems. Consequently, the crucial step is to show that the families f_ε and Df_ε, $\varepsilon\epsilon(0.1]$ are precompact in a suitably chosen topology; the related convergence to the corresponding term of (2.5). First we summarize the requirements this topology should satisfy.

(i) Since (2.5) is needed for all ρ_0, the subsequence we are going to select may not depend on λ, i.e. the measures $\mu_{\lambda,\varepsilon}$ may not be involved in the definition of the topology.

(ii) Since $\mu_{\lambda,\varepsilon}$ converges in a weak sense, see (1.19), these subsequences of f_ε and Df_ε must converge strongly, i.e. uniformly on some subsets B of H_e such that the integrals of (2.4) over the complement of such a B are uniformly small if B is large enough. To conclude that the limit of Df_ε equals Df, cf. (1.12), these B have to be convex.

(iii) In view of (i) and (ii), the Arzela-Ascoli Theorem will be used to select convergent subsequences of f_ε and of Df_ε in such a way that they converge uniformly on some compacts $B \subset H_e$. Since the strongly compact subsets of H_e are not fat enough to satisfy tightness bounds like $\mu_{\lambda,\varepsilon}(B) > 1-\delta$, uniformly in ε and λ, we need a weak topology of H_e.

(iv) To apply the Arzela-Ascoli Theorem, we have to verify that f_ε and Df_ε are equicontinuous. Moreover, we can pass from (2.4) to (2.5) only if we know (1.19) for all continuous and bounded functions, therefore this weak topology may not be too weak.

Taking (i)-(iv) into account, we will consider the usual weak topology of H_e, then the balls

$$B_a = [\sigma \varepsilon H_e : |\sigma|_r < a_r] \qquad (2.8)$$

are certainly compact; we have to prove that f_ε and Df_ε are uniformly bounded and weakly equicontinuous on such balls. Some tricks behind the related a priori bounds go back to [15,16], a detailed proof is given in [17] assuming that $d=1$ and $\alpha=0$.

The main tool of the proof of the priori bounds is the auxiliary vector process $\underline{Y}_\varepsilon = \underline{Y}_\varepsilon(t,x)$ defined by

$$d\underline{Y}_\varepsilon = \frac{1}{2}\, \nabla'(S_\varepsilon)\, dt - \frac{\alpha}{2}\, \varepsilon^2\, \nabla_\varepsilon \Delta_\varepsilon S_\varepsilon\, dt + d\underline{w}_\varepsilon \qquad (2.9)$$

and by $\underline{Y}_\varepsilon(0,x) = \underline{\omega}_\varepsilon(x)$, where $\underline{\omega}_\varepsilon \varepsilon L^d$ and $\nabla^*_\varepsilon \underline{\omega}_\varepsilon = \sigma_\varepsilon = S_\varepsilon(0,\cdot)$; then $\nabla^*_\varepsilon \underline{Y}_\varepsilon = S_\varepsilon$ holds for all $t > 0$, too. The nice, self-adjoint parabolic structure of $\underline{Y}_\varepsilon$ is the key to the following energy inequalities. Let E^σ denote the conditional expectation given the initial configuration σ. An easy calculation shows that, in view of (1.24) we have a universal constant C such that

$$\frac{d}{dt}\, E^\sigma\, |\underline{Y}_\varepsilon|_r^2 + c_1\, E^\sigma\, |S_\varepsilon|_r^2 < Cr + Cr^{-2}\, E^\sigma\, |\underline{Y}_\varepsilon|_r^2 , \qquad (2.10)$$

therefore, if $z > C$ then for all $\sigma \varepsilon H_e$ and $r > 1$ we have

$$c_1\, R_{z,\varepsilon} |\sigma|_r^2 < r + |K_\varepsilon \sigma|_r^2 , \qquad (2.11)$$

where $K_\varepsilon : H_e \to H_e\ L_\varepsilon$ denotes a distinguished inverse to ∇^*_ε, i.e. $\nabla^*_\varepsilon K_\varepsilon \sigma = I_\varepsilon \sigma$ is an identity. This inequality is sufficient to bound the resolvent $f_\varepsilon = R_{z,\varepsilon} g$ whenever g is Lipschitz continuous. To prove the equi-continuity of f_ε we have to compare solutions S_ε and $\bar{S}_\varepsilon$ with different initial configurations σ and $\bar{\sigma}$. Using the associated $\underline{Y}_\varepsilon$ and $\underline{\bar{Y}}_\varepsilon$ defined along the same Wiener trajectories, we obtain in a quite similar way that

$$c_1 R_{z,\varepsilon}|\sigma_\varepsilon - \bar{\sigma}_\varepsilon|_r^2 < |K_\varepsilon \bar{\sigma}|_r^2 \qquad \text{if} \quad z > C \ . \tag{2.12}$$

Since we can construct a unformly compact version of K_ε , the Lipschitz continuity of g is sufficient for the weak continuity of the resolvent $f_\varepsilon = R_{z,\varepsilon} g$.

The case of the functional derivative Df_ε is similar, but more involved. Differentiating (1.16) with respect to a parameter of the initial data, we obtain its first variational system

$$\frac{\partial u_\varepsilon}{\partial t} = \frac{1}{2} \Delta_\varepsilon(c_\varepsilon u_\varepsilon) - \frac{\alpha}{2} \varepsilon^2 \Delta_\varepsilon^2 u_\varepsilon \ , \tag{2.13}$$

where $u_\varepsilon = u_\varepsilon(t,y)$ and

$$c_\varepsilon = c_\varepsilon(t,y) = V''\,(S_\varepsilon(t,y)) \tag{2.14}$$

is bounded and it is bounded away from zero. If $S_\varepsilon(0,y) = \sigma_\varepsilon(y)$ then the functional derivative $Df_\varepsilon(\sigma,x)$ can be expressed in terms of the fundamental solution $p_\varepsilon = p_\varepsilon(t,x,y)$ to (2.13). For example, if $g(\sigma) = \phi(\sigma)$, $\phi \varepsilon H_e^*$, then

$$Df_\varepsilon(\sigma,x) = E^\sigma \int_0^\infty e^{-zt} \int p_\varepsilon(t,x,y)\phi(y)\, dy\, dt \ , \tag{2.15}$$

thus we must understand (2.13). The associated auxiliary process $\underline{v}_\varepsilon = \underline{v}_\varepsilon(t,y)$ is defined by

$$\frac{\partial \underline{v}_\varepsilon}{\partial t} = \frac{1}{2} \nabla_\varepsilon(c_\varepsilon u_\varepsilon) - \frac{\alpha}{2} \varepsilon^2 \nabla_\varepsilon \Delta_\varepsilon u_\varepsilon \tag{2.16}$$

with initial condition $\nabla_\varepsilon^* \underline{v}_\varepsilon(0,\cdot) = u_\varepsilon(0,\cdot)$; then $\nabla_\varepsilon^* \underline{v}_\varepsilon = u_\varepsilon$ for all $t > 0$, and the first energy inequality reads as

$$c_1 \int_0^\infty e^{-zt} \, |u_\varepsilon(t,.)|_r^2 \, dt < |\underline{v}_\varepsilon(0,.)|_r^2 \ , \quad z > C \ , \tag{2.17}$$

which implies the boundedness of Df_ε .

The equi-continuity property of Df_ε is obviously related to the continuous dependence of solutions to (2.13) on the coefficient c_ε . Indeed, if $\bar{c}_\varepsilon$ is another set of coefficients of type (2.14), and $\bar{u}_\varepsilon$ denotes the corresponding solution with $\bar{u}_\varepsilon(0,y) = u_\varepsilon(0,y)$, then using again the auxiliary processes $\underline{v}_\varepsilon$ and $\bar{\underline{v}}_\varepsilon$, i.e. differentiating $|\underline{v}_\varepsilon - \underline{v}_\varepsilon|^2$ with respect to time, a more or less direct calculation yields

$$c_1 \int_\delta^\infty e^{-zt} \, |u_\varepsilon(t,\cdot) - \bar{u}_\varepsilon(t,\cdot)|_r^2 \, dt \tag{2.18}$$

$$\leq C \int_0^\infty e^{-zt} \int \exp(\frac{|y|}{r}) \, |c_\varepsilon(t,y) - \bar{c}_\varepsilon(t,y)| \, u_\varepsilon^2(t,y) \, dy \, dt$$

for $z \geq C$. Here $c_\varepsilon - \bar{c}_\varepsilon$ is uniformly bounded, but it is small in a mean sense only, see (2.12), thus to conclude the equicontinuity of Df_ε , we need a moment condition that is stronger than (2.17), namely

$$\int_\delta^\infty e^{zt} \int |p_\varepsilon(t,x,y)|^{11/5} \exp(\frac{|y|}{2}) \, dy \, dt \tag{2.19}$$

$$\leq C_\delta \exp(\frac{|x|}{2r}) \quad \text{for } z \geq C \, , \ \delta > 0 \, .$$

The proof of (2.19) follows the perturbative treatment of parabolic equations by STOOCK-VARADHAN [34] , cf. FABES-RIVIERE [13] . This is the point where the restriction (2.1) is neded. A detailed proof of this result and some extensions are to be published in the Journal of Statistical Physics.

3. Concluding Remarks

1. The statement of the Theorem, which is essentially a law of large numbers, has been reduced to (1.19) by means of the resolvent equation method.

While (1.19) holds for weakly continuous functions only, from (2.12) we see that $f_\varepsilon = R_{z,\varepsilon} g$ and Df_ε are weakly continuous whenever g is Lipschitz continuous. On the other hand, if $g(\sigma) = \int \phi(x)h(\sigma(x))\,dx$ with some $\phi \varepsilon H_e^*$ and smooth $h : R \to R$ then (1.19) becomes

$$\int g(\sigma)\, \mu_{\lambda,\varepsilon}(d\sigma) \underset{\varepsilon \to 0}{\to} \int \phi(x) \int h(y)\, q_{\lambda(x)}(y)\, dy\, dx\ , \tag{3.1}$$

where q_u is the one-dimensional density of the homogeneous Gibbs state μ_λ with $\lambda_k = u$ for all k. Consequently, we can prove (1.4) - (1.6) by means of this modification of the resolvent method, whence the equation

$$\frac{\partial\lambda}{\partial t} = [F''(\lambda)]^{-1}\, \Delta\lambda$$

follows for the chemical potential $\lambda=\lambda(t,x)$. More generally,

$$E_{\lambda,\varepsilon}[\int \phi(x)h(S_\varepsilon(t,x))\,dx\,] \underset{\varepsilon \to 0}{\to} \tag{3.3}$$

$$\int \phi(x) \int h(y)\, q_{\lambda(t,x)}(y)\, dy\, dx\ ,$$

where $\lambda(t,x)$ is given by (3.2).

2. The above relation is still weaker than the usual form of the principle of local equilibrium. I believe that the nice entropy argument of ROST [31] reducing the problem to the law of large numbers, applies in this case, too.

3. Consider now the hydrodynamic version of (1.21), i.e. $a=2$, $\beta=1$, $b=0$ and the initial configuration is distributed by $\mu_{\lambda,\varepsilon}$ as before. In this situation the resolvent equation (2.4) turns into

$$\int g(\sigma)\, d\mu_{\lambda,\varepsilon} = z \int f_\varepsilon(\sigma)\, d\mu_{\lambda,\varepsilon} \tag{3.4}$$

$$+ \iint <\underline{J}(\sigma(x))\ ,\ \nabla_\varepsilon Df_\varepsilon(\sigma,x)> dx\, d\mu_{\lambda,\varepsilon}$$

$$-\frac{1}{2}\iint Df_{\varepsilon}(\sigma,x)\ \Delta_{\varepsilon}\lambda(x)\ dx\ d\mu_{\lambda,\varepsilon}\quad ,$$

which is very encouraging. Indeed, the energy inequalities (2.11),(2.12),(2.17) and (2.18) can be proven for (1.21) essentially in the same way as above, and (2.19) remains in force, too. Although the equicontinuity of the very singular $\nabla_{\varepsilon}Df_{\varepsilon}$ is not completely understood, we are in a position to conjecture that the density field of (1.21) satisfies the law of large numbers in the hydrodynamic limit, i.e. $S_{\varepsilon}(t,\phi) \to \int \phi(x)\rho(t,x)\ dx$ in probability, and ρ satisfies

$$\frac{\partial\rho}{\partial t} = \text{div}\ \underline{A}(\rho) + \frac{1}{2}\ \text{div}\ (\ D(\rho)\ \text{grad}\ \rho\ )\ , \tag{3.5}$$

where D is the same as above, while

$$\underline{A}(\rho) = \int \underline{J}(y)\ q_{\lambda}(y)\ dy \quad \text{if} \quad \rho = F'(\lambda)\ . \tag{3.6}$$

4. The technical condition (2.1) is relevant in the present proof, but it can certainly be relaxed, cf. FABES [12].

5. Perhaps the most important property of (1.16) we have exploited in the proof is the parabolicity of its first variational system (2.13); this property seems to be the right generalization of attractivity of $\underline{+}1$ spin systems.

4. Acknowledgements

I wish to express my thanks to H. Rost, H. Spohn, J.L. Lebowitz, G. Papanicolau, E. Fabes, R. Holley, M. Aizenman for useful comments and stimulating discussions. Parts of this work were done at I.M.A., University of Minnesota; I am indebted to G.R. Sell and H. Weinberger for their kind hospitality there.

<u>Note added in proof</u>: The proof, as outlined here, works only if $d = 1$. If $d > 1$ then the derivatives of the resolvent equation with respect to z, can be used.

References

1. Aizenman, M: Private communication, 1986

2. Andjel, E.: Invariant measures for the zero range process. Ann. Prob. 10. (1982) 525-547.

3. Boldrighini, C. - Dobrushin, R.L. - Suhov, Yu. M: The hydrodynamic limit of a degenerate model of classical statistical physics. Usp. Mat. Nauk 35.4. (1980) 152 (Short communication, in Russian)

4. Boldrighini, C. - Dobrushin, R.L. - Suhov, Yu. M: One-dimensional hard rod caricature of hydrodynamics. Journ. Stat. Phys 31. (1983) 577-616.

5. Brox, Th. - Rost, H: Equilibrium fluctuations of a stochastic particle system: The role of conserved quantities. Ann. Prob. 12. (1984) 742-759

6. De Masi, A. - Ianiro, N. - Pellegrinotti, S. - Presutti, E: A Survey of the Hydrodynamical Behaviour of Many-Particle Systems. Non-Equilibrium Phenomena II. Eds. Lebowitz, J.L. -Montroll, E.W., North-Holland Co. Amsterdam-New York 1984.

7. De Masi, A. - Ferrari, P.- Lebowitz, J.L.: Reaction-diffusion equations for interacting particle systems. Preprint 1986.

8. De Masi, A. - Presutti, E. - Vares, M.E.: Escape from the unstable equilibrium in a random process with infinitely many interacting particles. Preprint 1986.

9. Dobrushin, R.L: On the derivation of the equations of hydrodynamics, Lecture, Budapest 1978.

10. Dobrushin, R.L. - Siegmund-Schultze, R: The hydrodynamic limit for systems of particles with independent evolution. Math. Nachr. 105. (1982) 199-224.

11. Fabes, E: singular integrals and partial differential equations of parabolic type. Studia Math. XXVIII. (1966) 6-131.

12. Fabes, E: Private communication, 1986.

13. Fabes, E. - Riviere, N.M: Singular integrals with mixed homogenity. Studia Math. XXVII. (1966) 19-38.

14. Ferrari, P. - Presutti, E. - Vares, M.E: Hydrodynamics of a zero range model. Preprint 1984.

15. Fritz, J: Local stability and hydrodynamical limit of Spitzer's one-dimensional lattice model. Commun. Math Phys. 86. (1983) 363-373.

16. Fritz, J: On the asymptotic behaviour of Spitzer's model for the evolution of one-dimensional point systems. Journ. Stat. Phys. 38. (1985) 615-647.

17. Fritz, J: The Euler equation for the stochastic dynamics of a one-dimensional continuous spin system. Preprint 1986.

18. Fritz, J. -Gärtner,J: The continuum limit of a one-dimensional stochastic gradient system is a nonlinear stochastic partial differential equation. In preparation.

19. Funaki, T: Private communication, 1986.

20. Guo, M. - Papanicolau, G: Bulk diffusion for interacting Brownian particles. In: Statistical Physics and Dynamical Systems. 41-48. Eds. Fritz, J.- Jaffe, A. -Szász , D. Birkhäuser, Basel-Boston New-York 1985.

21. Hohenberg, P.C. - Halperin, B.I: Theory of dynamic critical phenomena. Rev. Mod. Physics 49. (1977) 435.

22 Kipnis, C. - Marchioro, C. - Presutti, E: Heat flow in an exactly solvable model. Journ. Stat. Phys. 22. (1982) 67.

23. Ma.S.K: Modern Theory of Critical Phenomena. Benjamin-Cummings, Reading, Mass. 1976.

24. Morrey, C.B.: On the derivation of the equations of hydrodynamics from statistical mechanics. Comm. Pure. Appl. Math. 8. (1955) 279-327.

25. Papanicolau, G. - Varadhan, S.R.S: Boundary value problems with rapidly oscillating coefficients. In: Random Fields Vol. II. 385-853, Eds. Fritz, J. - Lebowitz, J.L. - Szász, D, North Holland: Amsterdam-New York 1981.

26. Papanicolau, G. - Varahdan, S.R.S: Ohrstein-Uhlenbeck processes in a random potential. Commun. Pure. Appl. Math. XXXVIII. (1985) 819-334.

27. Presutti, E. - Scacciatelli, E: Time evolution of a one-dimensional point system: a note on Fritz's paper. Journ. Stat. Phys. 38. (1985) 64-654.

28. Rost, H: Non-equilibrium behaviour of a many-particle system: Density profile and local equilibrium. Z. Wahsch. verw. Geb. 58. 1981) 41.

29. Rost, H: Hydrodynamik gekoppelter Diffusionen: Fluktuationen in Gleichgewicht. In: Dynamics and Processes. Eds., Blanchard, Ph.- Streit, L., Lecture Notes in Mathematik 1031, 97-107, Springer Verlag: Berlin-Heidelberg-New York 1983.

30. Rost, H: Diffusion de sphéres dures dans la droite réelle: Comportement macroscopique et equilibre local. Lecture Notes in Mathematics 1059, 127-143, Eds. Azema, J. - Yor, M., Springer Verlag: Berlin-Heidelberg-New York 1984.

31. Rost, H: The Euler equation for the one-dimensional zero-range process. Lecture, Minneapolis, IMA, March 1986.

32. Spohn, H: Equilibrium fluctuations for some stochastic particle systems. In: Statistical Physics and Dynamical Systems 67-81. Eds. Fritz, J. - Jaffe, A. - Szász, D., Birkhaüser: Basel-Boston-Amsterdam 1985.

33. Spohn, H: Equilibrium fluctuations for interacting Brownian particles. Commun. Math. Phys. 103. (1966) 1-33.

34. Stroock, D.W. - Varadhan, S.R.S: Multidimensional Diffusion Processes. Springer Verlag: Berlin-Heidelberg-New York 1979.

CONVERGENCE OF THE RANDOM VORTEX METHOD

Jonathan Goodman

Courant Institute of Mathematical Sciences

New York University
New York, New York 10012

This will be rather informal and imprecise. For precise statements and proofs, see the references. I will first explain my motives and goals in studying the particle system that is the random vortex method. Then I will outline the proof of the convergence theorem. For more basic discussion of vortex methods, see the book [2]. Information on the practical success of vortex methods for computing viscous and inviscid flows can be found in [6]. Despite many published demonstrations of the effectiveness of vortex methods, the methods still have many vocal opponents. I hope this confusion will be clarified over the next few years, but I will not comment on it here. Recently, Long [7] has announced a different proof that leads to a sharper convergence theorem than the one discussed here.

The random vortex method is a numerical scheme for computing incompressible viscous flow in two or three space dimensions. It is intended for computing delicate physical phenomena that could be overwhelmed by numerical artifacts in standard finite difference calculations. It simulates the subtle combination of diffusion and advection that characterizes high Reynolds number (i.e. low viscosity) flow. However, it is not, in my view, a direct model of the physical flow; there is no diffusion oif physical "vortex particles" in real fluids. This becomes especially apparent if one tries to make up a vortex method for compressible flow or even when studying vortex methods for three dimensional flow. For these problems there is a great variety of possibilities with none of the choices seeming particularly more "physical" than the others. Therefore, we should develop methods using mathematical considerations of accuracy and stability and not restrict ourselves to "physical" models of a viscous fluid. In particular, we are willing to consider "smoothed vortex cores" rather than "point vortices". For Navier-Stokes equations, the "point" vortices" do not "exist" as physical entities, and their mathematical analysis is deep and difficult [8].

Smoothing is a central technical aspect of the random vortex method; it seems to add stability and accuracy. Without smoothing, the stability part of the proof below is almost certainly false. In particle simulations in plasma physics, an application with a much more direct physical basis, smoothing (the addition of "superparticles") is introduced for accuracy alone[3]. I might speculate that the improved accuracy in smoothed methods in plasma particle simulations is due to increased stability. Indeed, mild instabilities, such as instabilities in the unsmoothed vortex method, often show up as anomalously low accuracy in "stable" looking calculations. However, we do not want to add too much smoothing. We would like the "size of the vortex cores" to be on the order of the interparticle distance, or not too much bigger. The main shortcoming of my convergence proof is that it requires larger cores. More precisely, for N randomly placed points, the interparticle distance is on the order of $N^{-1/2}$ in two dimensions. Therefore, a "core size" $\varepsilon >> N^{-1/2}$ should be sufficient. I prove [4] that with high probability the computed velocity field is close to the true velocity field if $\varepsilon >> N^{-1/4}$.

The incompressible Navier Stokes equations in two space dimensions are

$$u_t + (u\cdot\nabla)u = \nu\nabla^2 u \,,$$
$$\mathrm{div}(u) = \nabla\cdot u = 0 \,.$$

Here, $u = (u_1,u_2)$ is the velocity and ν if the kinematic viscosity (see [2]). The vorticity ω is a scalar function defined by

$$\omega(x,t) = \mathrm{curl}(u)(x,t) = \nabla\times u = \frac{\partial u_1}{\partial x_2} - \frac{\partial u_2}{\partial x_1} \,.$$

The vorticity satisfies the simple advection diffusion equation

$$\omega_t + (\mu\cdot\nabla)\omega = \nu\nabla^2\omega \,. \tag{1}$$

The velocity can be determined from the vorticity by solving the linear first order elliptic system of differential equations $\nabla\cdot u = 0$, $\nabla \times u = \omega$. The solution of these equations is given by convolution,

$$u(\cdot,t) = K*\omega(\cdot,t) \quad , \tag{2}$$

where K is the "Biot-Savart kernel". K is given by

$$K(x) = \frac{c}{|x|^2}\cdot(x_2,-x_1) \quad ,$$

or

$$\hat{K}(\xi) = \frac{c}{|\xi|^2}(\xi_2,-\xi_1) \quad ,$$

where the hat represents the Fourier transform, $\hat{f}(\xi) = \frac{1}{4\pi^2}\int_{R^2} e^{-i\xi\cdot x}f(x)dx$.

The random vortex method is based on the observation that the vorticity transport equation(1) is also Kolmogorov forwards equation for the stochastic differential equation

$$dY(t) = u(Y(t),t) + \sqrt{\nu}dB(t) \quad . \tag{3}$$

It is convenient (but not essential) to pretend that $\omega(\cdot,t)$ is a Schwartz class probability density. In that case, given a sufficiently regular function $f(x)$, we have

$$E[f(Y(t)] = \int f(y)\omega(y)dy$$

If we had many independent particles, $Y_1(t),\ldots,Y_N(t)$ with $Y_j(0) \sim \omega(\cdot,0)$i.i.d., we could get a Monte Carlo approximation to $u(\cdot,t)$ using (2):

$$u(x,t) \approx v(x,t) = \frac{1}{N}\sum_{j=1}^{N} K(x-Y_j(t)) \quad . \tag{4}$$

Although this gives $E[v(x,t)] = u(x,t)$, we get low accuracy in any given sample because the variance of ν is infinite:

$$\mathrm{var}[v(x,t)] = \frac{1}{N}\cdot \mathrm{var}[K(x-Y_j(t))] = \frac{1}{N}\cdot\left(\int [K(x-y)-u(x,t)]^2\omega(y,t)dy\right) = \infty \quad .$$

The last integral has a logarithmic divergence for x-y small since

$$\int_{|x-y|\leq 1} |K(x-y)|^2 dy = \text{const.} \int_0^1 r^{-2} r dr = \infty \quad .$$

If we replace K by an approximate "desingularized" kernel in (4) then $E[\nu] \neq u$ but the error caused by a small bias can be more than compensated for by reduction in the error caused by noise. Let $\phi(x)$ be a Schwartz class function satisfying

$$\int \phi(x) dx = 1 \quad , \qquad \phi(x) = \phi(-x) \quad ,$$

and define $\phi_\varepsilon(x) = \varepsilon^{-2}\phi(\varepsilon^{-1}x)$, and $K_\varepsilon = \phi_\varepsilon * K$. Here ϕ describes the shape of a "vortex blob", ε is the "blob size", and K_ε is the desingularized (smoothed) kernel.

Let us assess the damage caused by regularizing. To do this we study the effect of replacing (2) by

$$u^\varepsilon(\cdot,t) = K_\varepsilon * \omega^\varepsilon(\cdot,t) \quad . \tag{5}$$

We are using the convention that an ε subscript implies convolution with ϕ_ε while an ε superscript indicates that the quantity depends on ε in some other way. The standard way to start is to show that ω, the exact solution of (1), (2), is an approximate solution to (1), (5). That is, ω satisfies,

$$\omega_t + u \cdot \nabla\omega = \nu\nabla^2\omega \quad ,$$

$$u = K_\varepsilon * \omega + \sigma^\varepsilon \quad ,$$

where σ^ε is quite small. From this it is routine (see section 2 of [4] to show that $\omega - \omega^\varepsilon$ is of the order of σ^ε. In Fourier transforms, σ^ε is given by

$$\hat{\sigma}^\varepsilon(\xi) = \hat{K}(\xi)\hat{\omega}(\xi)(1-\hat{\phi}(\varepsilon\xi)) \quad .$$

Therefore,

$$\|\sigma^\varepsilon\|_{L_2} = O(\varepsilon^\rho) \quad \text{if} \quad 1 - \hat{\phi}(\xi) = O(|\xi|^\rho) \quad \text{with} \quad \rho \geq 1 \quad .$$

This is equivalent to

$$\int x^\alpha \phi(x) dx = 0 \quad \text{if} \quad 1 \leq \alpha \leq \rho \quad ,$$

a moment condition first used in this context by Beale and Majda [1]. The conclusion is that smoothing errors can be minimized by careful choice of ϕ.

The continuous time random vortex method is then the coupled system of stochastic differential equations

$$dX_j(t) = v(X_j(t),t)dt + \sqrt{\nu}\, dB_j(t) \ , \tag{6}$$

$$v(x,t) = \frac{1}{N} \sum_{j=1}^{N} K_\varepsilon(x-X_j(t)) \ . \tag{7}$$

Let Y_j be i.i.d. solutions of (3) with B(t) replaced by i.i.d. noise $B_j(t)$. The above discussion should make plausible the claim that v will be close to u (i.e. the vortex method will be accurate) if the X_j are close to the corresponding Y_j. This approach (initiated by Hald [5] for vortex methods) is not advocated by people who study particle systems because it is too strong. We may have $u \approx v$ without pointwise agreement between the X_j and Y_j. The proof that X_j and Y_j are close proceeds in two steps as above. First we show that the Y_j almost satisfy the X_j equations (6),(7). That is

$$dY_j = \frac{1}{N} \sum_k K_\varepsilon(Y_j - Y_k)\, dt + \sqrt{\nu} dB_j + \rho_j dt \ ,$$

where

$$\rho_j = u(Y_j) - \frac{1}{N} \sum_k K_\varepsilon(Y_j - Y_k)$$

is small. This was discussed above.

The main step is stability. For our purposes it suffices to show linearized stability. That is, we may assume that ρ is infinitesimal and show that

$$\| X-Y \|_{l_2} = O(\rho) \ .$$

Here, if $z = z_1, \ldots, z_N$, then

$$\| z \|_{l_2}^2 = \frac{1}{N} \sum |z_j|^2 \ .$$

In our case we need only study the linearization of the mapping $Y \to v$ defined by

$$v_j = \frac{1}{N} \sum_k K_\varepsilon(Y_j - Y_k) .$$

The linearization is an $N \times N$ matrix whose (j,k) entry, A_{jk}, is a 2×2 block

$$A_{jk} = \frac{1}{N} \nabla K_\varepsilon(Y_j - Y_k) \qquad \text{if} \quad j \neq k ,$$

$$A_{jj} = \frac{1}{N} \sum_k \nabla K_\varepsilon (Y_j - Y_k) \quad ,$$

We will show that there is a Const independent of N so that with high probability $\|A\|_{\ell_2} < \text{Const}$.

This is done by looking at the N x N matrices B, C, and D, whose scalar entries are the components of the A_{jk} with $j \neq k$. (It is easy to treat the diagonal entries so we ignore them from now on.) From (3) we see that the entries of B, C, or D are the form $N^{-1} L_\varepsilon(Y_j - Y_k)$, where

$$\hat{L}_\varepsilon(\xi) = \frac{c\hat{\phi}(\varepsilon\xi)}{|\xi|^2} p_2(\xi)$$

and $p_2(\xi)$ is ξ_1^2, $\xi_1\xi_2$, or ξ_2^2, respectively. We will get a bound for the matrix $G = B^tB + C^tC + D^tD$, which implies bounds for the individual matrices B, C, and D. Consider, for example, $F = B^tB$. Apart from errors that are controllable, $F_{jk} = N^{-1} M_\varepsilon(Y_j - Y_k)$ where

$$M_\varepsilon = \int L_\varepsilon(x-y)\omega(y)L_\varepsilon(y)dy \quad . \tag{8}$$

The entries of F are given by

$$F_{jk} = \sum_\ell B_{j1} B_{k1}$$

$$= \frac{1}{N}\frac{1}{N} \sum_\ell L_\varepsilon(Y_j - Y_\ell) L_\varepsilon(Y_k - Y_\ell)$$

$$= \frac{1}{N} \cdot \text{(Monte Carlo approx. to (8))} .$$

The variance in this Monte Carlo approximation will be small provided that $\varepsilon >> N^{-1/4}$, a condition that plagues this proof. More smoothing is required here than earlier since L is more singular than K, L being pseudodifferential operator or order 0 while K has order -1. From (8) we see that the operator corresponding to M_ε is the product of three operators

$$M_\varepsilon = L_\varepsilon^t \cdot \omega \cdot L_\varepsilon ,$$

where L_ε corresponds to convolution with the kernel $L_\varepsilon(x)$ and ω represents pointwise multiplication by the function $\omega(x)$. From the calculus of pseudodifferential operators, we know that such operators commute up to operators of lower order. In particular,

$$M_\varepsilon = \omega \cdot L_\varepsilon^t \cdot L_\varepsilon + \text{op(ord=-1)} ,$$

where the second term on the right is negligible relative to the first. Since convolution in space goes over to multiplication of Fourier transforms, $L_\varepsilon \cdot L_\varepsilon$ has a kernel with transform

$$c\overline{\hat{\phi}}(\varepsilon\xi)\hat{\phi}(\varepsilon\xi) \cdot \frac{|p_2(\xi)|^2}{|\xi|^4}$$

The combination $G = B^tB + 2C^tC + D^tD$ has matrix elements corresponding to a kernel that (up to multiplication by ω and controllable errors) has transform

$$c|\hat{\phi}(\varepsilon\xi)|^2 \cdot \frac{\xi_1^4 + 2\xi_1^2\xi_2^2 + \xi_2^4}{|\xi|^4} = c(\hat{\phi}(\varepsilon\xi))^2$$

if $\hat{\phi}$ is real ($\hat{\phi}$ is real since ϕ is symmetric). The result of this is that we are reduced to showing boundedness for a matrix G with entries given by

$$G_{jk} = \frac{1}{N}\,\psi_\varepsilon(Y_j - Y_k) , \qquad \psi^\varepsilon = \phi_\varepsilon * \phi_\varepsilon$$

The final step is to apply a well known fact about matrices. Suppose H has bounded row and column sums:

$$\sum_j |H_{jk}| < C_c \quad \text{and} \quad \sum_j |H_{kj}| < C_r \quad , \tag{9}$$

then the l_2 norm of H is bounded by $\sqrt{C_c C_r}$. In the case of G, the two sums corresponding to (9) are both Monte Carlo approximations to

$$\int |\Psi^{\varepsilon}(x-Y_k)|\,\omega(x)dx \quad ,$$

so they will be bounded with high probability if ε is large enough that the variance is small.

References

[1] J.T. Beale and A. Majda, "Vortex methods II: Higher order accuracy in two and three dimensions", Math. Comp.39 (1982)pp.29-52

[2] A.J. Chorin and J. Marsden, A Mathematical Introduction to Fluid Mechanics, Springer Verlag, New York, 1979

[3] J.D. Dawson, "Particle simulations of plasmas", Rev. Mod. Phys., v. 55 (1983), pp. 403-447

[4] J. Goodman, "Convergence of the random vortex method", preprint

[5] A. Leonard, "Computing three-dimensional incompressible flows with vortex elements", in Annual Review of Fluid Mechanics, ed. by M. van Dyke, J.V. Wehausen, and J.L. Lumley, vol. 17 (1985), pp. 523-559

[6] O. Hald, "The convergence of vortex methods, II" SIAM J. Num. Anal., 16 (1979) pp. 762-755

[7] Long, PhD thesis, Princeton University, in preparation

[8] H. Osada, "Propagation of chaos for the two dimensional Navier-Stokes equation", preprint. See also these proceedings.

SUPERCRITICAL BRANCHING RANDOM FIELDS. ASYMPTOTICS OF A PROCESS INVOLVING THE PAST

Luis G. Gorostiza
Centro de Investigación y de Estudios Avanzados
Apartado Postal 14-740
México 07000, D.F., México

1. Introduction.

We consider a branching random field with immigration on R^d, described as follows. Initial particles appear at time $t = 0$ according to a Poisson random field with intensity measure γdx, $\gamma \geqq 0$, on a Borel set $B \subset R^d$. Immigrant particles appear according to a Poisson random field with intensity measure $\beta dxdt$, $\beta \geqq 0$, on a Borel set $C \subset R^d \times [0,\infty)$. The two Poisson fields are independent. As time evolves, the particles independently migrate following standard Brownian motions during independent exponentially distributed lifetimes with parameter V, at the end of which they independently reproduce with a branching law $\{p_n\}_{n=0}^{\infty}$ (i.e. a particle branches into n particles with probability p_n) having finite second moment; the mean and the second factorial moment of the branching law will be denoted m_1 and m_2, respectively. The offspring particles appear at the locations where their parent particles branched, and independently migrate, live and reproduce by the same laws. A more general model will be mentioned later on.

The usual process associated with this type of model is the point measure-valued process $\{N_t\}_t$ where $N_t(A)$ is the number of particles present in the Borel set $A \subset R^d$ at time t (e.g. [2,5,8,11]). However, this process contains no information on the past of the system; in particular it tells nothing about the locations of the ancestors of the particles observed at a given time. Our objective is to define and study a process which does convey this sort of information; such a formulation is useful when one wishes to focus attention on the particles whose ancestors were in given regions at certain times in the past.

Let $T > 0$ be the time when particles are observed, and for each $t \in [0,T]$ let $N_t^T(A)$ denote the number of particles present at time T such that their ancestors at time t had positions in the Borel set $A \subset R^d$. In order to account for all the particles we need to assume $p_0 = 0$, i.e. no particles vanish without descendants, and we remark that this assumption is consistent with the supercritical case only (i.e. $m_1 > 1$). It is not obvious how to carry out a similar analysis in the critical and subcritical cases.

We are interested in the asymptotic behavior of the measure-valued process $\{N_t^T\}_t$ as $T \to \infty$ under the space-time scaling $(x,t) \to (T^{\frac{1}{2}}x, Tt)$. In order to obtain meaningful results we must replace the immigration intensity β by $T^{-1}\beta$, and give to each particle a weight $e^{-\alpha T(1-s)}$ (after scaling), where $\alpha = V(m_1-1)$ is the Malthusian parameter of the underlying branching structure and s is the time of birth (before scaling) of the first ancestor of the particle. These weights are needed for the following reason. The random measure N_t^T has the same atom locations as the old random measure N_t, but now each atom has a random weight equal to the number of descendants of the corresponding particle at time T; therefore the total number of descendants at time T of all the atoms (particles) of N_t^T having a common first ancestor born at time s will grow as $e^{\alpha T(1-s)}$ as $T \to \infty$ (see [1]). We assume that the notation N_t^T already incorporates the scaling and changes just introduced, and we consider the process $N^T \equiv \{N_t^T, 0 \leqq t \leqq 1\}$.

We will give a law of large numbers and a functional central limit theorem for the process N^T as $T \to \infty$, as well as properties of the limit fluctuation process. The proofs appear in [9]. We will also comment on some aspects of the proofs that have bearing on the types of results we can prove, and we will mention another approach for the asymptotic analysis of this sort of model, which consists in describing the system by a random point-measure on function space.

2. Limit Theorems.

We denote $\{\mathcal{T}_t\}_t$ the standard Brownian semigroup (with $\mathcal{T}_t = 0$ for $t < 0$). $\mathcal{S}(R^d)$ and $\mathcal{S}'(R^d)$ designate the Schwartz spaces of rapidly decreasing C^∞ test functions and of tempered distributions, respectively, and $\langle\cdot,\cdot\rangle$ is the duality on these spaces. $\Rightarrow_f$ means weak convergence of finite-dimensional distributions of $\mathcal{S}'(R^d)$-valued processes (i.e. indexed by finitely many times and finitely many test functions).

We will assume that either (a) the sets B and $\mathcal{C} \cap (R^d \times [0,t])$ for each t have finite Lebesgue measures, or (b) $B = R^d$, $\mathcal{C}$ is of the form $\mathcal{C} = R^d \times C$ with Borel $C \subset [0,\infty)$, and $\{p_n\}$ has finite third moment.

The asymptotic behavior of N^T is given by the following results.

THEOREM 1 (*law of large numbers*). *For each* $0 \leqq t \leqq 1$ *and* $\phi \in \mathcal{S}(R^d)$,

$$T^{-d/2}\langle N_t^T,\phi\rangle \to \gamma\int_B \mathcal{T}_t\phi(x)dx + \beta\iint_{\mathcal{C}} \mathcal{T}_{t-r}\phi(x)dxdr$$

in L^2 *as* $T \to \infty$.

Let

$$M_t^T = T^{-d/4}(N_t^T - EN_t^T), \qquad 0 \leqq t \leqq 1,$$

denote the normalized fluctuation, and consider the process $M^T \equiv \{M_t^T, 0 \leqq t \leqq 1\}$.

THEOREM 2 (*functional central limit theorem*). $M^T \Rightarrow_f M$ *as* $T \to \infty$, *where*

$M \equiv \{M_t, 0 \leq t \leq 1\}$ *is a centered* $S'(R^d)$*-valued Gaussian process with covariance functional*

$$\mathrm{Cov}(\langle M_s,\phi\rangle,\langle M_t,\psi\rangle) = \frac{m_2}{m_1-1}\,[\gamma\int_B T_s\phi(x)T_t\psi(x)dx + \beta\iint_C T_{s-r}\phi(x)T_{t-r}\psi(x)dxdr].$$

The limit fluctuation process $M \equiv \{M_t\}$ in Theorem 2 is defined for all $t \geq 0$. We state now some of its properties (others are in [9]).

THEOREM 3.

(1) M *is Markovian and satisfies the generalized Langevin equation*

$$dM_t = \frac{1}{2}\,\Delta M_t dt + \left(\frac{m_2}{m_1-1}\right)^{\frac{1}{2}}\beta^{\frac{1}{2}}dW_t^C,$$

$$M_0 = \left(\frac{m_2}{m_1-1}\right)^{\frac{1}{2}}\gamma^{\frac{1}{2}}W^B,$$

where W^B *is standard Gaussian white noise on* B *and* $W^C \equiv \{W_t^C, t \geq 0\}$ *is a* $S'(R^d)$*-Wiener process with covariance functional*

$$\mathrm{Cov}(\langle W_s^C,\phi\rangle,\langle W_t^C,\psi\rangle) = \int_0^{s\wedge t}\int_{C_r}\phi(x)\psi(x)dxdr,$$

where $C_r = \{x \in R^d : (x,r) \in C\}$ *(i.e.* W^C *is standard space-time Gaussian white noise on* C*).*

(2) *In general* M_t *is not homogeneous (i.e. distribution invariant under spatial translations). If* $B = R^d$ *and* $C = R^d \times C$*, then* M_t *is homogeneous with spectral measure*

$$\sigma_t(d\lambda) = \frac{m_2}{m_1-1}\,[\gamma e^{-t\|\lambda\|^2} + \beta\int_{C\cap[0,t]}e^{-(t-r)\|\lambda\|^2}dr]d\lambda,$$

where $\|\cdot\|$ *is the Euclidean norm on* R^d.

(3) *When* $B = R^d$ *and* $C = R^d \times [0,\infty)$*, then* M *can be represented as*

$$M = \left(\frac{m_2}{m_1-1}\right)^{\frac{1}{2}}[\gamma^{\frac{1}{2}}M^I + \beta^{\frac{1}{2}}M^{II}],$$

where M^I *and* M^{II} *are independent processes which are self-similar under the transformations*

$$\langle a^{-d/2}M_{at}^I,\phi(\cdot/a^{\frac{1}{2}})\rangle \quad \text{and} \quad \langle a^{-d/2-1}M_{at}^{II},\phi(\cdot/a^{\frac{1}{2}})\rangle,$$

respectively, for any $a > 0$.

(4) *When* $B = R^d$, $C = R^d \times [0,\infty)$ *and* $d \geq 3$*, then* M_t *converges weakly as* $t \to \infty$

to a centered $S'(R^d)$*-valued Gaussian random variable* M_∞ *with covariance functional*

$$\mathrm{Cov}(\langle M_\infty,\phi\rangle,\langle M_\infty,\psi\rangle) = \frac{m_2}{m_1-1}\,\beta\,\frac{\Gamma(d/2-1)}{4\pi}\int_{R^d}\int_{R^d}\frac{\phi(x)\psi(y)}{\|x-y\|^{d-2}}\,dxdy,$$

where Γ *is the gamma function.* M_∞ *is homogeneous with spectral measure*

$$\sigma_\infty(d\lambda) = \frac{m_2}{m_1-1}\,\beta\,\frac{d\lambda}{\|\lambda\|^2}\,.$$

(Note that M_∞ *depends only on the immigration).*

Note: If $B = R^d$ and $C = R^d \times C$, then the law of large numbers takes the form

$$T^{-d/2}\langle N_t^T,\phi\rangle \to [\gamma+\beta\lambda_1(C\cap[0,t])]\int_{R^d}\phi(x)dx,$$

and

$$M_t^T = T^{-d/4}\{N_t^T - T^{d/2}[\gamma+\beta\lambda_1(C\cap[0,t])]\lambda_d\},$$

where λ_k denotes the Lebesgue measure on R^k.

3. Remarks.

(1) The model can be more general: the particle lifetimes are not necessarily exponentially distributed and the migration process may be a transport process which converges weakly under the space-time scaling. The limit theorems are similar but they involve additional effects and technicalities which we prefer to avoid here (see [4,6,9,10]).

(2) The reason for the two mutually exclusive assumptions (a) and (b) on the sets B and C depends on the fact that we have two different kinds of proofs. Case (a) is based on an almost-sure invariance principle for the empirical distribution of the trajectories of the descendance lines [10]; in this case we need the boundedness of the Lebesgue measures of B and $C\cap(R^d\times[0,t])$ in order to justify certain steps such as changing the order of limits and integrals. With this approach we can handle also the more general model (Remark (1)). Case (b) uses an explicit computation of the covariance functional of N^T which depends on the assumption $B = R^d$ and $C = R^d \times C$, and requires the finiteness of the third moment of the branching law in order to perform a limiting procedure. This approach does not cover the general model. (See [9]).

(3) The Langevin equation for the process M is obtained by an application of a general result on $S'(R^d)$-valued Gaussian processes [2].

(4) Note that in the case without immigration ($\beta = 0$) the process M obeys the heat equation with generalized random initial condition. This case was treated

[4] with $B = R^d$ and $C = R^d \times [0,\infty)$. In [4] the particular evolution equation for M was obtained by a special method [7]; the general result [2] was not yet known.

(5) We have not been able to prove tightness of the family $\{M^T\}_T$ in the Skorohod space $D([0,1],S'(R^d))$. In the case of B and C with finite Lebesgue measures and the general model (Remark (1)) we don't know how to proceed. In the special case considered here with $B = R^d$ and $C = R^d \times [0,\infty)$, we have a martingale for N^T and its increasing process ([9], Appendix). However, due to the complexity of these processes we have not been able to use them to prove tightness in the usual way (as in [8,11]).

4. A Measure on Function Space Approach.

We will consider here the case without immigration $(\beta = 0)$ for simplicity. Let $\tilde{N}^T(A)$ denote the number of descendance line trajectories of the system lying in the Borel set A of $C[0,1]$, the real-valued continuous functions on $[0,1]$, after the space-time scaling. For the same reason explained before we give to each atom of the random measure $\tilde{N}^T$ the weight $e^{-\alpha T}$, and we assume that the notation $\tilde{N}^T$ already incorporates these weights. Observe that if A is a cylinder with base Borel $A \subset R^d$ at time t, then $\tilde{N}^T(A) = N_t^T(A)$; thus we recover the measure-valued process $\{N_t^T\}_t$ from the random measure $\tilde{N}^T$ on $C[0,1]$. Now we define the fluctuation random (signed) measure on $C[0,1]$:

$$\tilde{M}^T(A) = T^{-d/4}(\tilde{N}^T(A) - E\tilde{N}^T(A)).$$

Let C_K denote the space of functions $\Phi: C[0,1] \to R^d$ which are bounded, continuous, and such that there is a compact set $K_\Phi \subset R^d$ such that $\Phi(u) = 0$ if $u(0) \notin K_\Phi$. Consider the characteristic functional of $\tilde{M}^T$:

$$F_{\tilde{M}^T}(\Phi) = E\exp\{i\int_{C[0,1]}\Phi d\tilde{M}^T\}, \qquad \Phi \in C_K.$$

Then it can be shown, using the almost-sure invariance principle [10], that

$$F_{\tilde{M}^T}(\Phi) \to F(\Phi) \equiv \exp\{-\frac{1}{2}\frac{m_2}{m_1-1}\int_{R^d}[\int_{C[0,1]}\Phi(u)\mathcal{B}^x(du)]^2 dx$$

as $T \to \infty$, where $\mathcal{B}^x$ is the standard Wiener measure on $C[0,1]$ with initial point x.

The first question this raises is: does there exist a generalized random Gaussian "object" $\tilde{M}$ on $C[0,1]$ whose characteristic functional is given by $F(\Phi)$ above? The answer is yes, and it can given by the stochastic integral

$$\tilde{M}(\Phi) = \left(\frac{m_2}{m_1-1}\right)^{\frac{1}{2}}\int_{R^d}\int_{C[0,1]}\Phi(u)\mathcal{B}^x(du)\tilde{W}(dx), \qquad \Phi \in C_K,$$

where $\tilde{W}$ is the standard Gaussian white noise measure on R^d (see e.g. [12]). The second question is: does the convergence of the characteristic functionals for each Φ imply the weak convergence of $\tilde{M}^T$ to $\tilde{M}$ in some sense? In order to answer this question we would like to have a nuclear Gelfand triple based on $C[0,1]$, in order to apply Lévy's continuity theorem for nuclear spaces [3]. This question remains unresolved.

This problem was posed in [6] and discussed in [4]. The present approach can also be used for the more general model (Remark (1)). There is a corresponding law of large numbers [4].

Acknowledgment

This research was partially supported by CONACyT grants PCCBBNA 002042 and 140102 G203-006 (Mexico) and a grant of the NSERC (Canada).

References

1. K.B. Athreya and P. Ney. "Branching Processes". Springer-Verlag, New York, 1973.
2. T. Bojdecki and L.G. Gorostiza. Langevin equations for S'-valued Gaussian processes and fluctuation limits of infinite particle systems. Z.f. Wahrschein. (to appear).
3. P. Boulicaut. Convergence cylindrique et convergence étroite d'une suite des probabilités de Radon. Z.f. Wahrschein., Vol. 28, (1973), 43-52.
4. D.A. Dawson and L.G. Gorostiza. Limit theorems for supercritical branching random fields. Math. Nachr., Vol. 118, (1984), 19-46.
5. D.A. Dawson and B.G. Ivanoff. Branching diffusions and random measures. In "Branching Processes", (A. Joffe and P. Ney, Eds.). Advances in Probability, Vol. 5, 61-103. Dekker, New York, 1978.
6. L.G. Gorostiza. Limites gaussiennes pour les champs aléatories ramifiés super-critiques. In "Aspects statistiques et aspects physiques des processus gaussiens", 385-398. Editions du CNRS, Paris, 1981.
7. L. G. Gorostiza. Generalized Gaussian random solutions of certain evolution equations. In "Advances in Filtering and Optimal Stochastic Control", (W.H. Fleming and L.G. Gorostiza, Eds.). Lecture Notes in Control and Inf. Sci., Vol. 42, (1982), 142-148.
8. L.G. Gorostiza. High density limit theorems for infinite systems of unscaled branching Brownian motions. Ann. Probab., Vol. 11, No. 2, (1982), 374-392.
9. L.G. Gorostiza. Limit theorems for supercritical branching random fields with immigration. Tech. Rep. No. 64. Lab. Res. Stat. Probab., Carleton Univ.-Univ. of Ottawa, 1985.
10. L.G. Gorostiza and N. Kaplan. Invariance principle for branching random motions. Bol. Soc. Mat. Mex., Vol. 25, No. 2, (1980), 63-86.
11. R. Holley and D.W. Stroock. Generalized Ornstein-Uhlenbeck processes and infinite particle branching Brownian motions. Publ. RIMS, Kyoto Univ., Vol. 14, (1981), 741-488.
12. E. Wong and M. Zakai. Martingales and stochastic integrals for processes with multidimensional parameter. Z.f. Wahrschein., Vol. 29, (1974), 109-122.

A SIMPLE MATHEMATICAL MODEL OF A SLURRY

David E. Loper
Department of Mathematics
Florida State University
Tallahassee, Florida 32306

and

Paul H. Roberts
Department of Mathematics
University of California
Los Angeles, California 90024

A slurry is defined here as a liquid containing solid particles in suspension. These particles are the solid phase which freezes from the liquid; they can melt or additional liquid can freeze as local conditions warrant. The absorption or release of latent heat as the particles melt or freeze causes the system to act as if it had an enhanced specific heat. Also the solid particles may rise or sink through the liquid under the action of gravity depending on whether they are lighter or heavier than the liquid. If solid forms at one level, releasing latent heat, and then migrates vertically and melts, absorbing latent heat, the net effect is a vertical transport of heat in the direction opposite to the motion of the solid. This is equivalent to the heat-pipe effect.

The liquid may be composed of more than one substance. In this case the solid is typically of a different composition than the liquid. Thus the density of the solid is governed by the composition of the system as well as the change of specific volume upon freezing. The vertical migration of solid can affect the local composition of the system. For example, if solid rich in one constituent migrates from a region, the effect is a local enhancement of all other constituents.

A general theory of a slurry was developed by Loper and Roberts (1978) using mixture theory; the slurry was considered to be a binary mixture of two chemically distinct constituents wherein at each point in space there are present both constituents in both phases in predictable proportions. The full nonequilibrium theory contains five independent thermodynamic variables. They also presented a reduced theory containing only three, relying on the simplifying assumptions of fast melting and constant solid composition. The former assumes that the timescales of interest are longer than that for relaxation to melting-freezing equilibrium while the latter assumes that the solid which forms is a pure substance, thus avoiding the problems of history dependence. This reduced theory was not Boussinesq; that is, the thermodynamic variables were not assumed constant and the nearly hydrostatic assumption was not made.

We report here on the development of a Boussinesq theory of a slurry starting from the general theory of Loper and Roberts (1978); for a more detailed exposition, see Loper and Roberts (1987). The theory rests on the assumptions of fast melting and constant solid composition plus the following five simplifying asumptions. (1). The flux of material in the liquid phase is independent of the temperature gradient (no Soret effect) and the flux of solid is independent of the composition and temperature gradients. The flux of solid depends only on the pressure gradient; this represents gravitational sedimentation. (2). The range of variation of the thermodynamic variables within the system are sufficiently small that the density and all thermodynamic coefficients are constant to dominant order. (3). The timescale of interest is long compared with the passage of a sound wave across the system. (4). All dissipative heating terms are negligibly small. (5). The system is to a good approximation in hydrostatic equilibrium.

The result of these simplifications is the following set of equations governing the variation of the mass-averaged velocity, $\mathbf{u}$, temperature, T, perturbation pressure, p, and mass fraction of light constituent, ξ:

$$\nabla \cdot \mathbf{u} = 0,$$

$$D\xi/Dt = [\delta^* g(\partial b/\partial\phi)]_0 \Big\{[\xi/(1-\phi)]_0 (\partial\xi/\partial z) + [(1-\phi)L/T\mu]_0 (\partial T/\partial z) +$$

$$[(1-\phi)\delta\rho g/\mu]_0\Big\} - [D^*/T]_0 \nabla^2 T,$$

$$D\mathbf{u}/Dt = -\nabla p/\rho_0 + g_0[\alpha_0^* (T-T_0) + \rho_0\delta_0^* (\xi-\xi_0)]\mathbf{z} + \nu_0 \nabla^2 \mathbf{u},$$

$$DT/Dt + [\alpha Tg/C_p{}^*]_o\, \mathbf{u}\cdot\mathbf{z} = \kappa_o{}^* \nabla^2 T$$

where $D/Dt = \partial/\partial t + \mathbf{u}\cdot\nabla$ is the motional derivative, z is the vertical coordinate, ϕ is the mass fraction of solid, ρ is the density, g is the acceleration of gravity, L is the latent heat of fusion, ν is the kinematic viscosity, δ is the change of specific volume upon melting at constant composition, μ is the change of chemical potential with composition, b is the sedimentation coefficient,

$$\alpha^* = \alpha + (1-\phi)^3 \rho\delta L/\mu\xi^3$$

is and effective coefficient of thermal expansion (α),

$$D^* = (1-\phi)^2 DL/\mu\xi$$

is an effective material diffusivity (D),

$$\kappa^* = [kT\mu\xi^2 + (1-\phi)^3 \rho DL^2]/[\rho C_p T\mu\xi^2 + \rho(1-\phi)^3 L^2]$$

is an effective thermal diffusivity (k is the thermal conductivity and C_p is the specific heat),

$$\delta^* = \delta + (1-\phi)\delta/\xi$$

is an effective change of specific volume with composition (δ), and a subscript o denotes a constant value. The effective coefficients are changed from their normal values by the effect of the change of phase; see Loper and Roberts (1978).

The set of governing equations possesses a static equilibrium solution of the form

$$\mathbf{u} = \mathbf{0}, \qquad T = T_o + \eta(z-z_o), \qquad \xi = \xi_o - \zeta(z-z_o),$$

$$p = [\alpha^*\eta + \rho_o\delta_o{}^*\zeta]\rho_o g_o (z-z_o)^2/2,$$

where the thermal and compositional gradients are assumed to be related by

$$[L/T]_o \eta - [\mu\xi/(1-\phi)^2]_o \zeta + [\rho\delta g]_o = 0$$

so that the mass fraction of solid is independent of depth.

In the absence of buoyancy effects, the static solution has two stable perturbation modes representing viscous and thermal decay plus a neutrally stable mode which involves the effects of melting-freezing and sedimentation. When buoyancy effects are added, this static state is found to be absolutely unstable. For details, see Loper and Roberts (1987).

References

Loper, D. E., and Roberts, P. H., "On the motion of an iron-alloy core containing a slurry I. General theory," Geophys. Astrophys. Fluid Dyn., 9, 289-321 (1978).

Loper, D. E., and Roberts, P. H., "A Bousinesq model of a slurry," in Structure and Dynamics of Partially Solidified Systems, D. E. Loper, ed., Martinus-Nijhoff, xxx-xxx (1987).

LIMIT POINTS OF EMPIRICAL DISTRIBUTIONS

OF VORTICIES WITH SMALL VISCOSITY

Hirofumi Osada
Department of Mathematics
Nara Women's University
Kita-Uoya Nishimachi, Nara 630
JAPAN

§Introduction

Let $v(t,z)$ $(z=(x,y)\varepsilon R^2)$ be the vorticity of an incompressible and viscous two dimensional fluid, under the action of an external conservative field. Then v is described by the following evolution equation

$$(0.1) \qquad \partial_t v + (u\cdot\Delta)v - \nu\Delta v = 0, \qquad u(t,z) = (\Delta^{\perp}G)*v(t,z),$$

where $G(z) = -(2\pi)^{-1}\log|z|$, * denotes convolution, $\Delta = (\frac{\partial}{\partial x}, \frac{\partial}{\partial y})$ and $\Delta^{\perp} = (\frac{\partial}{\partial y}, -\frac{\partial}{\partial x})$. Here $\nu > 0$ denotes the viscosity constant. As far as strong solutions concerns, (0,1) is equivalent to the Navier-Stokes equation. In fact, $u(t,z)$ turns to be the velocity field described by the Navier-Stokes equation. Conversely we can get v from u as v = curl u. Since the two dimensional Navier-Stokes equation is an equation of a vector valued function, a probabilistic treatment is not easy, while the vorticity equation (0.1) is nothing but a McKean's type non-linear equation (see [3]. Such an observation for the two dimensional Navier-Stokes equation was made by Marchioro-Pulvirenti in [2].

Let $\{Z_t\}$ denote the McKean process associated with (0.1);

$$(0.2) \qquad dZ_t = \sigma dB_t + u(t,Z_t)dt \qquad u(t,z) = (\Delta^{\perp}G)*(Z_t\circ P)(z)$$

where $\sigma^2=2\nu$, $\{B_t\}$ is a 2-dimensional Brownian motion.

The n particle system associated with (0.1) are described by the following SDEs,

$$(0.3) \qquad \{ \; dZ_t^i = \sigma dB_t^i + (n-1)^{-1}\sum_{\substack{j\neq i\\ j=1}}^{n} (\Delta^{\perp}G)(Z_t^i - Z_t^j)dt, \qquad 1\leq i\leq n,$$

where $(B_t^1,\ldots,B_t^n)$ is a 2n-dimensional Brownian motion. Since the coefficients of

(0.3) have singularities at

$N = \bigcup_{i \neq j}^{n} \{z=(z_1,\dots,z_n) \in R^{2n}, z_i \neq z_j\}$, it is not trivial to see that the solution of (0.3) defines a conservative diffusion process on R^{2n}. However, if it starts out side of N, it can be shown that this diffusion process does not hit N Osada [6].

We prepare some notations. MCS, denotes the probàbilities on (S,B(S)) for a separable metric space S, and $< m, f > = \int_S f\, dm$.

C always denotes $C\{[0,\infty) \to R^2\}$. $Pn \in M(C^n)$ (resp. $P \in M(C)$) denote the distribution of the solution of (0.3) ((0.2)) with an initial distribution

$\phi(Z_1)\ \phi(Z_2)\ \dots\dots\ \phi(Z_n)\ dz_1\ \dots\dots\ dz_n\ (\phi(z)dZ)$. Define for k=1,2,...

$$\overline{Z}_{n,k} = \frac{1}{I(n,k)^{\#}} \sum_{I(n,k)} \delta\,(Z^{i1}_{\cdot}, \dots\dots, Z^{ik}_{\cdot})$$

with $I(n,k) - \{(i_1,\dots i_k);\ 1 \leq i_p \leq n,\ i_p \neq i_q$ if $P \neq q\}$, and let $\overline{P}n,k \in M(M(C^k))$ be the distribution of $\overline{Z}_{n,k} \circ Pn$.

Our problem is to show

$$(0.4) \qquad \lim_{n\to\infty} \overline{P}n.k = \delta \underbrace{p\ x\dots x\ p}_{k} \quad \text{in} \quad M(M(C^k)).$$

The importance of this problem lies in the fact aht, it (0.4) is valid for some k, then the propagation of chaos is valid, that is

$$(0.5) \qquad \lim_{n\to\infty} Pn,m = \underbrace{P\ x\dots x\ P}_{m} \quad \text{in } M(C^m) \quad \text{for all } m$$

where $Pn,m = (Z^1_t,\dots,Z^m_t) \circ Pn$ and P =s the distribvution of (0.2).

Several works on this problem have been done. Marchioro-Pulvirenti[2] and Goodman [1] discussed this problem under cut-off assumptions on the drift term. It was proved that

Theorem 1 ([7']) There exist a positive constant $\nu o\ (> \frac{1}{2\pi})$ such that if $\nu > \nu o$, then

$$\lim_{n\to\infty} Pn,m = P \times \dots \times P = n\ M(C^m)$$

for all m and $\phi \in L^\infty (R^2)$.

Theorem 1 needs the assumption that ν is large. However, as for the tightness of $\{ \overline{P}n,m \}$, we have

Theorem 2. ([7'] $\{\overline{P}n,m\}$ n=1,2... is tight for all m and $\nu > 0$.

Hence we denote by $\overline{P}_\infty$ an arbitrary limit point of $\{\overline{P}n,2\}$ and by $\{\overline{P}n\}$ a convergent subsequence of $\{ \overline{P}n.2 \}$ to $\overline{P}_\infty$. The purpose of this paper is to show for a.e.m.w.r.t.$\overline{P}\infty$, Z_t^1 om a "weak" solution of (0.1) for all ν greater than zero.

Let $\mu \in M(c)$. We say μ is a W_1 - solution of (0.1) if $\mu_t = Z_t o\mu$ satisfies the following conditions:

$(W_1\text{-}1)$ $\int_0^t\int_{R^4} |z_1-z_2|^{-1} d\mu_s(dz_1)d\mu_s(dz_2)ds < \infty$,

$$<\mu_s,\phi(s,\cdot)>|_{s=0}^{s=t} + \int_0^t <\mu_s,-\partial_s\phi(s,\cdot) - \nu\Delta\phi(s,\cdot)> ds$$

$$- \int_0^t <\mu_s \times \mu_s, (\Delta^{\perp}G)(z_1-z_2)\cdot(\Delta\phi)(z_1) > ds = 0$$

for all $\phi \in C_0^2([0,\infty)xR^2)$ and $0 < t < \infty$.

Moreover, we say μ is a W_0 - solution of (0.1) if $\mu_t = Z_t o\mu$ satisfies the following conditions:

$(W_0\text{-}1)$ $\mu_t \times \mu_t (\{(Z_1,Z_2) : Z_1 = Z_2\}) = 0$ a.e. t ,

$(W_0\text{-}2)$ $<\mu_s, \phi(is,.)>|_{s=0}^{s=t} + \int_0^t < \mu_s, - \partial_s\phi(s,.) - \nu\Delta\phi(cs,.)> ds$

$$- \int_0^t <\mu_s \times \mu_s, H > ds = 0,$$

where $H (t,Z_1, Z_2) = \frac{1}{2} (\Delta^{\perp}G)(Z_1-Z_2) \cdot (\Delta\phi(t,Z_1) - \Delta\phi(t,Z_2))$

It should be noted that $H \in L^\infty ([0,\infty) \times R^2 \times R^2)$ because

$$| H(t,Z_1,Z_2) | \leq \frac{1}{4\pi} \cdot |Z_1-Z_2|^{-1} |\Delta\phi(t,Z_1) - \Delta\phi(t_1z_2)|,$$

and that the expectation of $(W_0\text{-}2)$ is well defined by means of $(W_0\text{-}1)$.

Since $\Delta^{\perp}G(Z_2-Z_1) = -\Delta^{\perp}G(Z_1-Z_2)$, it is clear that μ is a W_1-solution if and only if μ is a W_0-solution satisfying $(W_1\text{-}1)$.

Now we state our main result

Theorem. For $\overline{P}_\infty$ a.ν.m, $\hat{m} = Z^1_{\cdot} \circ m$ is a W_0-solution of (0.1) for all $\nu > 0$.

Let $\overline{P}_{\infty,k}$ be an arbitrary limit point of $\{\overline{P}_{n,k}\}$ n=k,k+1,..., Then, as a corollary of Theorem, we have

Theorem[1] For $\overline{P}_{\infty,k}$ a.s.m ε $M(C^k)$, $\hat{m} = Z^1_{\cdot} \circ m$ is a W_0-solution of (0.1) for all $\nu > 0$, and all k = 1,2....

The author would like to express his sincere thanks to Professors S. Nakao and S. Kotani for their useful comments on Lemma 1.

§1. Proof of Theorem

Let $S(f,\lambda)$ denote the SDE defined by

$$(1.1)\qquad \begin{aligned} dZ^i_t &= dB^i_t + \frac{1}{n-1}\sum_{\substack{j\neq i\\ j=1}} (\Delta^{\perp}f)(Z^i_t - Z^j_t)\,dt\,, \\ Z_0 &\sim \lambda \qquad\qquad (\lambda \in M(R^{2n})). \end{aligned}$$

where $f(z)$ is a function on R^2. $S(f,z)$ denotes $S(f,\delta_z)$.

Let $\{G_\rho\}$ denote a sequence of smooth functions on R^2 with

$$(1.2)\qquad \lim_{\rho\to 0} G_\rho = G \text{ compact uniformly on } R^2 - \{(0,0)\}$$

Lemma 1. Let $Q_\rho \in M(C^n)$ denote the distribution of the solution of $S(G_\rho,\delta_z)$. Then

$$(1.3)\qquad \lim_{\rho\to 0} Q_\rho = Q \text{ in } M(C^n) \text{ for all } z \in R^{2n} - N\,,$$

where $N = \bigcup_{i \neq j}^{n} \{ (Z_1,\ldots,Z_n) \in R^{2n};\ Z_i = Z_j \}$ and

Q is the distribution of the solution of $S(G,Z)$.

Proof. Let $G^{\varepsilon}(z)$ (resp. $G^{\varepsilon}_{\rho}(z)$)

be a smooth function on R^2 with $G^{\varepsilon}(z) = G(z)$ $(G^{\varepsilon}_{\rho}(z) = G_{\rho}(z))$

if $|z| > \varepsilon$. We assume

$$(1.4) \qquad \lim_{\rho \to 0} G^{\varepsilon}_{\rho} = G^{\varepsilon} \quad \text{uniformly in } R^2 .$$

Let $\{Z^{\rho,\varepsilon}_t(w)\}$ (resp. $\{Z^{\varepsilon}_t(w)\}$ be the strong solution of $S(G^{\varepsilon}_{\rho},z)$ $(S(G^{\varepsilon},z))$.
Then by (1.4), we have

$$(1.5) \quad \lim_{\rho \to 0} Z^{\rho,\varepsilon}_t(w) = Z^{\varepsilon}_t(w) \quad \text{uniformly in } t \text{ in } [0,T] \text{ a.s.w.}$$

Let τ be the function on $C^n = C\{[0,\infty,\to R^{2n}\}$ defined by

$$\tau(\xi) = \inf\{t;\ 0 < t \leq T,\ \xi_t \in N_{\varepsilon}\} \qquad \text{for } \xi = (\xi_t) \in C^n ,$$

where $N_{\varepsilon} = \bigcup_{i \neq j}^{n} \{(z_1,\ldots,z_n)\ ;\ |z_i - z_j| < \varepsilon\}$. Then, it is clear that,

for a.s.w, τ is continuous at $\{Z_t^{\rho\varepsilon}(w)\}$. Hence we have by (1.5) that

$$(1.6) \qquad \lim_{\rho \to 0} \tau_{\rho,\varepsilon}(w) = \tau_{\varepsilon}(w) \qquad \text{a.s.w,}$$

where $\tau_{\rho,\varepsilon}(w) = \tau(\{Z^{\rho,\varepsilon}_t(w)\})$ and $\tau_{\varepsilon}(w) = (\{Z^{\varepsilon}_t(w)\})$.
We have by (1.5) and (1.6) that

$$(1.7) \qquad \lim_{\rho \to 0} Z^{\rho,\varepsilon}_{\tau_{\rho,\varepsilon}} = Z^{\varepsilon}_{\tau_{\varepsilon}} \qquad \text{a.s.w.} -$$

Since $Z^{\rho,\varepsilon}_{\tau\rho\varepsilon} = Z^{\rho,0}_{\tau\rho,0}$ and $Z^{\varepsilon}_{\tau\varepsilon} = Z_{t(z(w))}$ if $z \in R^{2n} - N$,
(1.7) yields

$$(1.8) \qquad \lim_{\rho \to 0} Z^{\rho,0}_{\tau\rho,0} = Z_{\tau}(z(w)) \quad \text{for all } 0 < \varepsilon < \infty \text{ and } z \in R^{2}n - N\varepsilon.$$

Here $\{Z_t\}$ is the solution of $S(G,Z)$. One can show (see [6]) that such a $\{Z_t\}$ exists uniquely and $P_z\ \{\inf\ \{0<t;\ Z_t \ \varepsilon\ N\} < \infty\} = 0$. Hence (1.8) concludes (1.3).

We prepare analytical estimates.

Lemma 2. Let C_{ij} be smooth functions on R^n with $|C=j| \leq d/n$ and $\sum_{i=j=1}^{n} \partial i \partial_j C_{ij} = 0$. Let $A = \nu\Delta + \sum_{j=1}^{n} (\Delta_i C_{ij})\ \Delta_j$.

Then the fundamental solution of $\partial_t - A$ satisfies the following:

(i) $P(t,x,y) \leq c_1 t^{-n/2}\ e^{-c_2|x-y|^2/t}$ for all $0<t<\infty$ and $x.y\ \varepsilon R^n$, where c_1, c_2 are positive constants depending only on ν, α and n,

(ii) $|\ P(t',x',y') - P(t,x,y)| \leq C_3\ \{1t-t'1^{\theta/2} + 1x-x'1^{\theta} + 1y-y'1^{\theta}\}$

for all $0<T \leq t,t'$, $\chi,x'y,y'\ \varepsilon\ sR^n$, where C_3 is a positive constant depending only on T, ν, α and n , and θ $(0<\theta<1)$ depends only on ν, α and n,

(iii) $\int_{R^n} P(t,x,y) \sum_{i=1}^{n} |x_i - y_i|^k\ dy \leq C_4\ nt^{k/2}$ for all $t > 0$,

where C_4 is a positive constant depending only on k, ν and α .
See [8] for the proof of (i) and (ii) , and Lemma 1.3 in [7'] for (iii).

Let $g_\rho(z) = (2\pi\rho)^{-1}\ e^{-|z|^2/2\rho}$ and $G_\rho = g_\rho * G$ (* denotes convolution). Let P_n^ρ denote the distribution of the solution of $S(G_\rho, x_n\ \phi dz)$.

Let $P\rho(t,z,z')$ be the fundamental solution of $\partial_t - {}_\rho$, where ${}_\rho = \nu\Delta - \frac{1}{n-1} \sum_{\substack{i\neq j\\ ij=1}}^{n} (\Delta^{\perp} G_\rho(z-z_j)\Delta_j$, which is the formal adjoint of the generator of $(Z_t,\ P_n^\rho)$.

Lemma 3.

(1.9) $< P_n^\rho\ ,\ |Z_t^1 - Z_s^1\ 1 > \ < \ C_5(t-s)^{1/2}$ for all $0<t-s<\infty$, with a positive constant C_5 depending only on ν,

(1.10) $\quad P_\rho(t,z,z') \leq c_0\, t^{-n/2} \exp(-c_7 |z-z/|^2/t)$

with positive constants C_0 and C_7 depending only on ν and n.

(1.11) $\quad \lim_{\rho\to 0} P_\rho(+.z,z') = P(t,z,z') \quad$ a.e. (t,z,z').

Proof L^1_ρ satisfies conditions in Lemma 2 with $d = 2$. Indeed

$$\partial_x G = \partial x a_1 + \partial_y a_2$$
$$\partial_y G = \partial x a_3 + \partial_y a_1$$

with $\quad a_1(z) = -x^2y^2/\pi|z|^4 , \quad a_2(z) = -3xy/2\pi|z|^2 + x^3y / \pi|z|^4$

$$a_3(z) = -3xy/2\pi|z|^2 + y^3x/\pi|z|^4.$$

Hence $\quad \partial_x G_\rho = \partial_x a_1 * g_\rho + \partial y\, a_2 * g_\rho,\ \partial_y G_\rho = \partial_x a_3 * g_\rho + \partial_y a_1 * g_\rho .$

We have $\quad |a_i * g_\rho| \leq 3/4\pi$, which implies $\alpha=2$.

(1.9) and (1.10) follow from Lemma 2. (0.11) follows from Lemma 1 and 2 immediately.

Let $G^+ = \max\{0, G\}$ and $G^- = \min\{0,G\}$.

Lemma 4.

$$\overline{\lim_{n\to\infty}} \ < Pn, \int_0^t G^+ (Z_s^1 - Z_s^2) ds > \ < \infty .$$

Proof. We first observe

$$\Delta G_\rho (z_1) \cdot \Delta^1 G_\rho(Z_2) = 0$$

and that

$$\Delta G_\rho \leq 0 .$$

Hence we have by Ito's formula that

$$< P_n^\rho , G_p (Z_t^1 - Z_t^2) > < P_n^\rho , G_\rho(Z_0^1$$

By (1.10) and (1.11) we can apply Lebesque's convergence theorem to obtain

$$< Pn, G(Z_t^1 - Z_t^1) > \leq < Pn, G(Z_0^1 - Z_0^2) > .$$

Then

$$< Pn, G^+(Z_t^1 - Z_t^2) > \leq < Pn, \{G^- (Z_o^1) - G^- (Z_t^1 - Z_t^2)\} >$$
$$+ < Pn, G^+ (Z_o^1 - Z_o^2) >$$
$$\leq C < Pn, |Z_t^1 - Z_o^1| > + < Pn, G^+ (Z_o^1 - Z_o^1) >$$

This together with (1.9) and $\phi \in L^\infty (R^2)$ concludes Lemma 4.

Lemma 5.

$$\overline{\lim_{n\to\infty}} < \overline{P}n, < m, \int_o^t G^+ (Z_s^1 - Z_s^2) ds >> < \infty .$$

Proof. Lemma 5 is an immediate consequence of

$$< \overline{P}n, < m, \int_o^t G^+ (Z_s^1 - Z_s^2) ds >> = < Pn, \int_o^t G^+ (Z_s^1 - Z_s^2) ds >$$

and Lemma 4.

Lemma 6 (Sznitman [10]). For a.s.m. w.r.t. $\overline{P}_\infty$

$m = \hat{m} \times \hat{m}$ with $\hat{m} = Z^1_{\cdot} \circ m$

<u>Proof of Theorem</u> By Lemma 5 and 6, we have for a.s.m w.r.t. $\overline{P}_\infty$, $\tilde{m} = Z^1_{\cdot} o m$ satisfies $(W_o - 1)$.

Let H^+ (resp. H^-) be the upper (lower) semicontinuos version of $H(t_1\ z_1, Z_2) = 1/2\ (\Delta^{\perp}G)(Z_1 - Z_2) \cdot (\Delta\phi(t,z) - \Delta\phi(t,Z_2))$.

Let $F^\pm$ be the function on $M(C^2)$ defined by

$$F^\pm(m) = < m, \phi(t_1\ Z_t^1) - \phi(0,Z_o^1) - \int_o^t(\partial_s\phi + \nu\Delta\phi)(s,Z_s^1)ds >$$
$$- m, \int_o^t H^\pm (s,Z_s^1 , Z_s^2) > .$$

We set

$$F_1^\pm = \max \{F^\pm, 0\} \quad \text{and} \quad F_2^\pm = \min \{F^\pm , 0\} .$$

By Ito's formula, $F^{\pm}(m) = 0$ for a.s.m. w.r.t. $\overline{P}n$. Then

$$< \overline{P}_n, F_1^{\pm} > = < \overline{P}_n, F_2^{\pm} > = 0$$

Then, since $F_2^{\pm}$ is upper semicontinuous, we have

$$0 \geqslant < \overline{P}_\infty, F_2^{+} \geqslant \overline{\lim_{n\to\infty}} < \overline{P}_n, F_2^{+} > = 0 .$$

Similarly, we have

$$0 \leqslant < \overline{P}_\infty , F_1^{-} > \leqslant \underline{\lim}_{n\to\infty} < \overline{P}n, F_1 > = 0.$$

By $(W_0\text{-}1)$, we have $< \overline{P}_\infty, F_2^{+} > = < \overline{P}_\infty, F_2^{-} >$. Hence

we concludes

$$< \overline{P}_\infty, |F^{-}| > = < \overline{P}_\infty, F_1^{-} > - < \overline{P}_\infty, F_2^{-} > = 0,$$

which shows that for a.s.m w..r.t. $\overline{P}_\infty$, $\hat{m}$ satisfies $(w_0 - 2)$.

References

[1] J. Goodman, Convergence of the random vortex method (to appear).

[2] C. Marchioro and M. Pulvirenti, Hydrodynamic in two dimensions and vortex theory, Commun. Math. Phys. 84, (1982) 483-503.

[3] H.P. McKean, Propagation of chaos for a class of nonlinear equation, Lecture series in differential equations, Session 7, Catholic Univ., (1967).

[4] T. Miyakawa, Y. Giga and H. Osada, The two dimensional Navier-Stokes flow with measures as initial vorticity, to appear.

[5] H. Osada, Moment estimates for parabolic equations in the divergence form, J. Math, Kyoto Univ. 25-3 (1985) 473-488.

[6] H. Osada, A stochastic differential equation arising from the vortex problem, Proc. Japan Acad., 61, Ser. A (1985).

[7] H. Osada, Propagation of chaos for the two dimensional Navier-Stokes equation, Proc. Japan Acad., 62, Ser. A (1986) (announcement).

[7′] H. Osada, Propagation of chaos for the two dimensional Navier-Stokes equation, to appear.

[8] H. Osada, Diffusion processes associated with generalized divergence forms, to appear.

[9] D.W. Stroock and S.R.S. Varadhan, Multidimensional diffusion Processes, Springer-Verlag.

[10] A.S. Sznitman, Propagation of chaos result for the Burgers equation, to appear.

[11] A.S. Sznitman, Nonlinear reflecting diffusion process and the propagation of chaos and fluctuations associated, J. Func. Anal., 56 (1984), 311-339.

[12] H. Tanaka, some probabilitistic problems in the spatially homogenous Boltzmann equation, Proc. of IFIP-ISI conf. on theory and applications of random fields, Bangalore, (1982).

MATHEMATICAL STUDY OF SPECTRA IN RANDOM MEDIA

Shin Ozawa

Department of Mathematics
Osaka University
Toyonaka, Osaka 560
Japan

The Lenz shift phenomena concerning eigenvalues of the Laplacian in random media was studied by various authors by various methods. In this note we give an expository introduction to this research area.

Let M be a bounded region in R^3 with smooth boundary γ. Let β be a fixed number > 1. We remove $[m^\beta]$ balls of centers $w_1,\ldots,w_{\tilde{m}}$ ($= w(m)$), $\tilde{m} = [m^\beta]$ with the same radius α/m (α = const.) from M.

Let $\mu_j(w(m))$ be the j-th eigenvalue of the Laplacian in $M \setminus \tilde{m}$-balls under the Dirichlet condition on its boundary.

In this note, we make an expository introduction to a study of statistical properties of $\mu_j(w(m))$ when we set a statistical law on all configuration of the centers of balls $w(m) \in M^{\tilde{m}}$. Behaviour of $\mu_j(w(m))$ as $m \to \infty$ is discussed.

Fix $\beta \in [1,3)$ and $\alpha > 0$. We consider M as a probability space by fixing a positive continuous function V on M satisfying

$$\int_M V(x)dx = 1$$

so that

$$P(x \in A) = \int_A V(x)dx.$$

Let $M^{\tilde{m}}$ be the product probability space. All configuration of the centers of balls $w(m)$ can be considered as a probability space $M^{\tilde{m}}$ by the statistical law stated above. Hereafter $\mu_j(w(m))$ is considered as a random variable on $M^{\tilde{m}}$.

Rigorous mathematical study of the above problem was given in 1974 by M. Kac [5] and Huruslov-Marchenko [4] . They obtained the following:

(I) The Lenz shift phenomena

Assume $\beta = 1$. Fix V. Fix j. Then, $\mu_j(w(m)) - \mu_j(V)$ tends to zero in probability as $m \to \infty$. Here $\mu_j(V)$ is the j-th eigenvalue of $-\Delta + 4\pi\alpha V(x)$ in M under the Dirichlet condition on γ.

The notion and the theory of Wiener sausage was used to give (I) when $V(x) \equiv$ const. ($= |M|^{-1}$). The method in [4] is potential theoretic and is not perturbative. Rauch-Taylor [16] gave (I) using Feynman-Kac formula and Wiener sausage. Papanicolaou-Varadhan [15] studied a similar problem concerning heat equation by probabilistic argument. See, also recent research by Chavel-Feldman [2], Anne [1].

Another approach to (I) was settled by [9], [11]. This method is perturbative. Notice Kac's opinion on perturbative method in page 525 of [5].

We want to make a heuristic argument to prove (I). Before to state it, we show a result.

We consider a fixed domain in R^3. Assume that $0 \in D$ and $R^3 \setminus \overline{D}$ is connected. Let w be a fixed point in M. Put $D_\varepsilon = \{x \in R^3 ; \varepsilon^{-1}(x - w) \in D\}$ and $M_\varepsilon = M \setminus \overline{D}_\varepsilon$. Let $\mu_j(\varepsilon)$ (respectively, μ_j) be the j-th eigenvalue of $-\Delta$ in M_ε (respectively, M) under the Dirichlet condition on M_ε (respectively, γ).

We have the following result : (Ozawa [8]).

(II) Capacity plays an important role.

Fix j. assume that μ_j is simple. Then,

$$\mu_j(\varepsilon) = \mu_j + 4\pi\varepsilon \ \mathrm{Cap}(D)\phi_j(w)^2 + O(\varepsilon^{2-s})$$

holds for any $s > 0$. Here ϕ_j denotes the normalized eigenfuction associated with μ_j. Here Cap(D) denotes the electrostatic capacity of the set D with respect to the point at infinity ∞. Note that Cap(D) = 1 when D is a unit ball. We know $\mathrm{Cap}(D_\varepsilon) = \varepsilon\, \mathrm{Cap}(D)$.

Heuristic argument on the Lenz shift using (II).

Assume that $V(x) \equiv |M|^{-1}$ (= const.) in (I). We have

(III) $$\mu_j(w(m)) = \mu_j + (4\pi\alpha/m)\ \mathrm{Cap}\ (D) \sum_{i=1}^{m} \phi_j(w_i)^2$$

$$+ R(w_1, \dots, ,w_m).$$

If we can say that the measure of the set satisfying

(IV) $$R(w_1, \dots, w_m) = O(m^{-t})$$

for $t > 0$, tends to 1 as $m \to \infty$, then

$$\mu_j(w(m)) - (\mu_j + 4\pi\alpha \mathrm{Cap}(D) \cdot |M|^{-1}) \to 0$$

in probability by law of large numbers.

Note also Ozawa [7] in which we can find heuristic.

We know that everything is in (IV). Here the author should emphasize that heuristic argument must be modified when $V(x) \neq |M|^{-1}$. However, the above observation is very helpful to study our problem.

Rigorous study of the Lenz shifts leads to :

Result 1 (Ozawa [11],[12]). Fix $\beta \in [1,3)$. Fix j. Then, there exists $\delta > 0$ such that

$$m^{-(\beta-1)+\delta} (\mu_j(w(m)) - u_j(m;V))$$

tends to zero in probability as $m \to \infty$. Here $u_j(m;V)$ denotes the j-th eigenvalue of $-\Delta + 4\pi\alpha m^{\beta-1}V(x)$ in M under the Dirichlet condition on γ.

Remark. Thus, $-\Delta$ in $M \setminus m$ balls $\to m^{\beta-1}(-m^{1-\beta}\Delta + 4\pi\alpha V(x))$ in M as $m \to \infty$ in a sense. It has a connection with semi-classical approximation of Schrödinger operator. The case $\beta > 1$ was first examined by the author.

Result 2. (Fluctuation of spectra, Figari-Orlandi-Teta [3; $\beta = 1$] , Ozawa [10; $\beta \in [1,12/11)$, $V \equiv |M|^{-1}$]) .

Fix j and β in the above interval. Assume that $V \equiv |M|^{-1}$, if $\beta > 1$. Assume that the j-th eigenvalue $\mu_j(m;V)$ (for $\beta > 1$, $\mu_j(m;V) = \mu_j + 4\pi\alpha m^{\beta-1}|M|^{-1}$ of $-\Delta + 4\pi\alpha m^{\beta-1}$ v in M under the Dirichlet condition on γ is simple. Then, a random variable $m^{1-(\beta/2)}(\mu_j(\cdot) - \mu_j(m;V))$ tends in distribution to a Gaussian random variable

$$\pi_j \text{ of mean } E(\pi_j) = 0 \text{ and variance } E(\pi_j^2)$$

$$= 4\pi\alpha(\int_M \phi_j^V(x)^4 V(x)dx \quad - \quad (\int_M \phi_j^V(x)^2 V(x)dx)^2$$

as $m \to \infty$. Here ϕ_j^V is the normalized eigenfunction associated with $u_j(m;V)$.

Remark . Result 2 on fluctuation is also observed by a simple modification of (III). If we have a good remainder estimate of $R(w_1,\ldots,w_m)$, then we see

$$m^{1/2}((4\pi\alpha/m) \sum_{i=1}^{m} \phi_j(w_i)^2 - 4\pi\alpha|M|^{-1}) \to \pi_j$$

as $m \to \infty$ (when V = const.). Thus, Result 2 is nothing but the central limit theorem for sums of independent identically distributed random variables modulo error term.

The following question naturally arises.

Question : Can one get the same result for higher β ?

The answer is no when $\beta = 3$. See [10]. Here the author offers the following conjecture (as a working hypothesis).

Conjecture. (Transition of fluctuation)

Let I_n denote the interval $[3(n-1)/n, 3(n/n+1))$, $(n = 2,3,\ldots)$.

For $\beta \in [1,2)$ the term $\mu_j(\cdot) - \mu_j(m;V)$ has the form $m^{(\beta/2)-1}\pi_j$ + lower term as $m \to \infty$.

At $\beta = 2$ there occurs a phenomena of transition of fluctuation. More precisely and daring to say, ...

For $\beta \in I_n$ $(n = 3,4,\ldots)$ the term $\mu_j(\cdot) - \mu_j(m;V)$ has the form

$$m^{\sigma(n;\beta)}\pi_j(n) + \text{lower term}$$

as $m \to \infty$. Here $\sigma(n;\beta)$ is a linear function of β in I_n. There is a relation

$$\partial\sigma(n;\beta)/\partial\beta|_{\beta\in I_n} < \partial\sigma(n+1;\beta)/\partial\beta|_{\beta\in I_{n+1}}$$

The relation

$$\lim_{n\to\infty}\max_{\beta\in I_n}\sigma(n;\beta)|_{\beta\in I_n} = 2$$

holds.

<u>Clustering property of obstacles suggests the above conjecture</u> : We say that

$(w_{i_1}^{(1)},\ldots,w_{i_{s(1)}}^{(1)})$, $(w_{i_1}^{(2)},\ldots,w_{i_{s(2)}}^{(2)})\ldots,(w_{i_1}^{(p)},\ldots,w_{i_{s(p)}}^{(p)})$ forms clusters, when

$$|w_k^{(t)} - w_q^{(t)}| < m^{-1}\log m \qquad (k,g = 1,\ldots,i_{s(t)})$$

for $\quad t = 1,\ldots,p,$ and

$$\min_{k,q}|w_k^{(t)} - w_q^{(s)}| > m^{-1+\xi}\ (t \neq s)$$

for some $\xi > 0$. We call $\max s(k)$ maximal size of clusters.

(V): It is easy to see that the measure of the set such that maximal size of clusters is less than $n + 1$ tends to 1 as $m \to \infty$ for $\beta\in I_n$

<u>Proof.</u> Assume that $\beta \in I_n$. Assume that there exists $w_{i_1},\ldots,w_{i_{n+1}}$ such that $|w_{i_k} - w_{i_q}| < m^{-1}\log m$. Then,

$$P\,(\text{maximal size of clusters} \geq n+1)$$

$$\leq (m^{\beta})^{n+1}(m^{-1}\log m)^{3n}.$$

We get the desired result.

Consider the case where $(w_1,\ldots,w_n)$ is in a cluster. Moreover we assume that $|w_i - w_j| \sim c_{ij}m^{-1}$. Then, (III) should be corrected as follows :

(VI) :

$$\mu_j(w(m)) - \mu_j(m;V) =$$

$$= 4\pi \operatorname{Cap}\,(\bigcup_{i=1}^{n} B(w_i))\phi_j(\tilde{w})^2 + 4\pi\,(\alpha/m) \sum_{i=n+1} \phi_j(w_i)^2$$

$$+ \ldots .$$

Here $B(w_i)$ denotes the ball of the center w_i with radius α/m. Here $\tilde{w}$ denotes one of w_i $(i = 1,\ldots,n)$.

Recall that

(VII):

$$\operatorname{Cap}(\bigcup_{i=1}^{n} B(w_i)) < \sum_{i=1}^{n} \operatorname{Cap}(B(w_i)) = 4\pi n\alpha m^{-1},$$

when w_i $(i=1,\ldots,n)$ is in a cluster

By observing (V), (VI), (VIII), the author made the above conjecture. The conjecture is in other words "Transition of fluctuation of spectra is brought from local fluctuation of capacity of obstacles". Even though the above conjecture fails, we will have a progress by checking the validity of the above slogan.

Strategy : (See [10]).

Approximation of Green's function is a key to (II), (III). The eigenvalue problem of the Laplacian is transformed into the eigenvalue problem of the Green operator. Primitive idea of our proof of (II), (III) is a construction of an approximate Green function of $-\Delta + Tm^{\beta-1}$ $(T = \text{const.} > 0)$ in $M \setminus$ balls under the Dirichlet condition on $\partial(M \setminus \text{balls})$ by using the Green function of $-\Delta + TM^{\beta-1}$ in M under the Dirichlet condition on γ. It should be noticed that $M \setminus$ balls may not be connected. We can avoid this technical complication by the following restriction $\mathcal{O}_1(m)$ on $w(m) \in M^{\tilde{m}}$ which assures that the only one connected component ω plays a role of $M \setminus$ balls and that the other components are negligible to study (II). $\mathcal{O}_1(m)$: Take an arbitrary open ball K of radius $m^{-\beta/3}$ in R^3. Then, the number of balls such that $\overline{\text{ball}}$ $K \neq \phi$ is at most $(\log m)^2$.

We see that P ($\mathcal{O}_1(m)$ holds) tends to 1 as $m \to \infty$. The eigenvalue problem of $-\Delta + Tm^{\beta-1}$ in ω is transformed into the eigenvalue problem of the Green operator. Let $G(x,y)$ be the Green function of $-\Delta + \lambda$ $(\lambda = Tm^{\beta-1})$ in M under the Dirichlet condition. Let $G(x,y;w(m))$ be the Green function of $-\Delta + \lambda$ in ω under the Dirichlet condition.

Let G ($G_{w(m)}$, resp.) be the bounded linear operator on $L^2(m)$ $(L^2(\omega)$, resp.) defined by the integral kernel $G(x,y)$ $(G(x,y;w(m))$, resp.). We examine the eigenvalue problem of $G_{w(m)}$. As a limit $m \to \infty$, we know that $\mu_j(w(m)) + \lambda$ is approximated by the j-th eigenvalue of $-\Delta + \lambda + 4\pi\alpha m^{\beta-1}V$ in M under the Dirichlet condition.

A key idea is an introduction of operators $H_{w(m)}$ and $\tilde{H}_{w(m)}$. The following integral kernel $h(x,y;w(m))$ was introduced in [10] (See also [11]). We abbreviate $G(w_i,w_j)$ as G_{ij}. Put $q = \exp(\lambda^{1/2}\alpha/m)$. Put $m^* = (\log m)^2$.

$$h(x,y;w(m)) = G(x,y) + (-4\pi\alpha/m)\, q \sum_{i=1}^{\tilde{m}} G(x,w_i)G(w_i,Y)$$

$$+ \sum_{s=2}^{m^*} (-4\pi\alpha/m)^s q^s \quad \sum\nolimits_{(s)} G(x,w_{i_1})G_{i_1 i_2}\cdots G_{i_{s-1} i_s} G(w_{i_s},Y).$$

Here the indices in $\sum_{(s)}$ run over all $1 \leq i_1 \ldots i_s \leq \tilde{m}$ such that $i_k \neq i_p$ when $k \neq p$. The sum $\sum_{(s)}$ is a <u>self-avoiding sum.</u>

Put

$$(H_{w(m)}f)(x) = \int_\omega h(x,y;w(m))f(y)dy, \qquad x \in \omega,$$

and

$$(H_{w(m)}g)(x) = \int_M h(x,y;w(m))g(y)dy, \qquad x \in M.$$

We compare $G_{w(m)}$ with $H_{w(m)}$, $H_{w(m)}$ with $\tilde{H}_{w(m)}$ and $\tilde{H}_{w(m)}$ with A. We see that $H_{w(m)}$ is a nice approximation of $G_{w(m)}$ under some restriction $\mathcal{O}_2(m)$ on $w(m)$. Before to state $\mathcal{O}_2(m)$, we should say that $G_{w(m)} - H_{w(m)}$ and $H_{w(m)} - \tilde{H}_{w(m)}$ are negligible to study fluctuation and that fluctuation comes from $\tilde{H}_{w(m)} - A$. An analysis of fluctuation coming from $\tilde{H}_{w(m)} - A$ is a contribution of [3].

It should be remarked that we need estimates of the operator norms

$$\|G_{w(m)} - H_{w(m)}\|_{L^2(\omega)},$$

$\|\chi_\omega H_w(m)\chi_\omega - \tilde{H}_{w(m)}\|_{L^2(M)}$, and $\|\tilde{H}_{w(m)} - A\|_{L^2(M)}$. Here χ_ω is a characteristic function of ω. The analysis of $L^2(M)$ norm of $(\chi_\omega H_{w(m)}\chi_\omega - \tilde{H}_{w(m)})f$ for fixed f does not suffice the investigation of spectra.

This approximation scheme can be understood by manipulating various terminology in multiple scattering formalism in physics. See, for example Ziman [19]. Some physisist may think that our rigorous investigation is out of question. I will ask ...

"Can you easily see that $H_{w(m)}$ approximates $G_{w(m)}$ very well ?"

"..."

"No ! A priori we did not know that the Dirichlet boundary condition can be replaced by delta (δ) potential (in Born expansion of Green's function). Even if we know it, we need extremely hard calculation to justify our intuition. Our mathematical investigation makes a firm basis of understanding a piece of multiple scattering theory. Moreover, we can proceed further and further..." See [18]

Approximation scheme does not work when $\{w_i\}$ forms a bad configuration.

We must clarify sufficient condition on $w(m)$ which can be used to estimate $H_{w(m)} - G_{w(m)}$. We introduce $\mathcal{O}_2(m)$. See [10].

$\mathcal{O}_2(m)$: There exists a constant c independent of m such that

$$\sum_{(s+1)} |w_{i_1}-w_{i_2}|^{-2+\sigma} \exp(-\lambda^{1/2}|w_{i_1}-w_{i_2}|)$$
$$\times G_{i_2 i_3} \cdots G_{i_s i_{s+1}}$$
$$\leq C^{s+1}\lambda^{-s+(1/2)(1-\sigma)} m^{\beta(s+1)} q(m)$$

holds for any $1 \leq s \leq (\log m)^2$, $\sigma = 0,1$. Here $q(m)$ denotes a function of m satisfying $q(m)/(\log m)^2 \to \infty$ as $m \to \infty$.

We see that $P(\mathcal{O}_2(m)$ holds) tends to 1 as $m \to \infty$. See [10] .

We can estimate $\|G_{w(m)} - H_{w(m)}\|_{L^2(\omega)}$ under the assumptions $\mathcal{O}_1(m), \mathcal{O}_2(m)$. And we see that the difference $G_{w(m)} - H_{w(m)}$ is negligible to consider fluctuation.

It should be remarked that $\mathcal{O}_1(m)$ and $\mathcal{O}_2(m)$ are mild so that obstacles can intersect each other.

Related topics

It is interesting to the author to apply our calculation to the problem of interacting particle systems (of course, we need some modifications). Especially, our analysis will be a strong guide to further development on the problem examined in Lang-Nguyen-Xaun [6].

There is a result on the eigenvalues with small Neumann obstacle (reflecting boundary). See Ozawa [13]

Our calculation using Greens function can be applied to various operators such as biharmonic operator, Laplace-Beltrami operator, etc. See, Rubinstein [17] in which Stokes operator case is discussed.

A development our our study (including higher β) will be discussed in [4].

Acknowledgement

The author would like to express his sincere gratitude to the staff of Institute for Mathematics and its Applications for their hospitality while he was there. He also thanks to Professor G.C. Papanicolaou for his constant support on the study.

References

[1] C. Anne.: Probleme de la glace pilee. Séminaire de Théorie spectrale et Géométrie. Chambery-Grenoble 1984-1985.

[2] I. Chavel, E.A. Feldman.: The Riemannian Wiener sausage. preprint.

[3] R. Figari, E. Orlandi, S. Teta.: The Laplacian in regions with many obstacles. Fluctuations around the limit operator. J. Stat. Physics. 41, (1985), 465-487.

[4] E. Ja. Huruslov, V.A. Marchenko.: Boundary value problems in regions with fine grained boundaries. (in Russian) Kiev 1974.

[5] M. Kac.: Probabilistic methods in some problems of scattering theory. Rocky Mountain J. Math. 4 (1974), 511-538.

[6] R. Lang, X.Nguyen-Xuan.: Smoluchowski's theory of coagulation in colloids holds rigorously in the Boltzmann-Grad limit. Colloquia Mathematica Societatis János Bolyai. 27. Random Fields, Esztergom (Hungary) 1979.

[7] S. Ozawa.: Singular variation of domains and eigenvalues of the Laplacian. Duke Math. J., 48, (1981), 767-778.

[8] S. Ozawa.: Electrostatic capacity and eigenvalues of the Laplacian. J. J. Fac. Sci. Univ. Tokyo Sec.IA, 30, (1983), 53-62., announcement, Proc. Japan Acad. 58, (1982), 134-136.

[9] S. Ozawa.: On an elaboration of M. Kac's theorem concerning eigenvalues of the Laplacian in a region with randomly distributed small obstacles. Commun. Math. Phys. 94, (1983), 473-487.

[10] S. Ozawa.: Fluctuation of spectra in random media. to appear in Proc. Probabilistic Methods in Math. Phys. held at Katata, Japan.

[11] S. Ozawa.: Random media and eigenvalues of the Laplacian. Commun. Math. Phys. 94, (1984), 421-437.

[12] S. Ozawa.: Spectral properties of random media. Proc. Japan Acad. 60A, (1984), 343-344.

[13] S. Ozawa.: Spectra of domains with small spherical Neumann boundary. J. Fac. Sci. Univ. Tokyo. Sec. IA, 30, (1983), 53-62., announcement, Proc. Japan Acad. 58, (1982), 134-136.

[14] S. Ozawa.: Approximation of Green's function in a region with many obstacles. in preparation 1986.

[15] G. C. Papanicolaou, S. R. S. Varadhan.: Diffusion in region with many small holes. Lect. Notes in Control and Information. vol 75, Springer (1980).

[16] J. Rauch, M. Taylor.: Potential and scattering theory on wildly perturbed domains. J. Funct. Anal., 18, (1975), 27-59.

[17] J. Rubinstein.: On the macroscopic description of slow viscous flow past a random array of spheres. preprint.

[18] which may be given by someone

[19] J. M. Zaiman.: Models of Disorder. Cambridge Univ. Press. 1979.

HYDRODYNAMIC SCREENING IN RANDOM MEDIA

Jacob Rubinstein

Institute for Mathematics and its Applications
University of Minnesota
Minneapolis, MN 55455

and

Department of Mathematics
Stanford University
Stanford, CA 94305

I. Introduction

We study the propagation of momentum through viscous fluid in a domain containing a large number of randomly distributed small obstacles. The obstacles are assumed to be fixed, and the velocity is governed by Stokes equations. Our motivations are:

(i) Analyse hydrodynamic interactions in random configurations.

(ii) Develope models for porous media.

Several scalings are compared and a uniform transition from Stokes equation to Darcy law is discovered. Diffusion problems in a similar set up were treated by many authors ([3], [9], [10] and [11], among others). Some physical background and formal multiple scattering expansions are provided in [1], [2], [4], [5], [13] and [15]. The mathematical theory of Stokes equation is extensively studied in [7] and [16].

II. Formulation and Preliminaries

Let D be a bounded domain in R^3 (possibly R^3 itself), $\{\mathbf{y}_j\}$ a collection of N fixed points in D which are randomly distributed with a probability density function $\rho(\mathbf{x})$, and let $\{B_j^N\}$ be N spheres of radius R^N centered at $\{\mathbf{y}_j\}$. We define

$$D^N = \{\mathbf{x} \in D;\ \mathbf{x} \in \bigcap_1^N |\mathbf{x} - \mathbf{y}_j| > R^N\},$$

and we consider in D^N the Stokes semigroup which is generated by

$$\begin{aligned} &\frac{\partial \mathbf{u}^N}{\partial t} = \mu\Delta\mathbf{u}^N - \nabla P^N + \mathbf{F}(\mathbf{x}) \\ &\nabla\cdot\mathbf{u}^N = 0 \end{aligned} \qquad \text{in } D^N \qquad (2.1)$$

$$\mathbf{u}^N|_{S_j^N} = 0, \quad \mathbf{u}^N|_{\partial D} = 0, \quad S_j^N = \partial B_j^N$$

$$\mathbf{u}^N(\mathbf{x},0) = \mathbf{u}_0^N(\mathbf{x}).$$

u^N is the velocity, μ is the viscosity, p^N the pressure and F is a given body force. We are interested in approximating $\mathbf{u}^N$ as N becomes large by a smooth vector field $\mathbf{v}$. To this purpose we will compare several scalings of the form

$$R^N = \frac{\alpha(N)}{N} = \frac{\alpha_0 N^\gamma}{N} \qquad \gamma \in [0,\gamma_0), \qquad (2.2)$$

where γ_0 will be determined later on.

The "natural" scaling is $\gamma = 0$. To motivate this statement, recall that the drag on a single sphere of radius R in a uniform flow with velocity w at ∞ is $6\pi\mu Rw$. If we assume that there are N noninteracting spheres, then the total drag on the system is $6\pi\mu NRW$, and the choice $R = O(\frac{1}{N})$ makes it an $O(1)$ quantity. As we shall see, the corresponding macroscopic equation is

$$\begin{aligned} &\frac{\partial \mathbf{v}}{\partial t} = \mu\Delta\mathbf{v} - 6\pi\mu\rho(\mathbf{x})\ \alpha(N)\mathbf{v} - \nabla P + \mathbf{F} \qquad (2.3) \\ &\nabla\cdot\mathbf{v} = 0 \end{aligned} \qquad \text{in } D$$

$$\mathbf{v}\ |_{\partial D} = 0, \quad \mathbf{v}(\mathbf{x},0) = \mathbf{u}_0(\mathbf{x}).$$

It is more convenient to analyse (2.1) and (2.3) through the resolvents, and we write the resolvent equation (without changing notation):

$$\mu\Delta\mathbf{u}^N - \lambda\mathbf{u}^N = \nabla P^N - \mathbf{F}\ , \qquad (2.4)$$

and

$$\mu\Delta\mathbf{v} - \lambda\mathbf{v} - 6\pi\mu\alpha\rho\mathbf{v} = \nabla P - \mathbf{F} \tag{2.5}$$

The case $\gamma = 0$ was studied in [12] and the main result was

Theorem 2.1 ($\gamma = 0$)

Assume that ρ, $\mathbf{F} \in C_0(D)$. Then

$$\lim_{N\to\infty} P_{\Omega^N}\{N^\sigma \|\mathbf{u}^N - \mathbf{v}\|_{L_2} < \varepsilon\} = 1 \tag{2.6}$$

$$\forall\ \varepsilon > 0,\quad \lambda \geq \lambda_0 \quad \text{and} \quad \sigma < \frac{1}{6}\ .$$

Here Ω^N is the space of configurations, and P_{Ω^N} is the probability measure induced there by ρ.

The main purpose of this paper is to extend Theorem 2.1 to $\gamma > 0$. In what follows we assume that D is a smooth compact domain. (The case $D = R^3$ can be analysed by the same methods.) Let us recall (without proofs) a few basic propositions which were discussed in [12]:

Lemma 2.1

Let $\Omega_1^N \subseteq \Omega^N$ be the set of all configurations that satisfy:

$$\min_{i\neq j} |\mathbf{y}_i - \mathbf{y}_j| \geq CN^{-1+\nu} \qquad \nu < \frac{1}{3} \tag{2.7}$$

$$N^{-2} \sum_{i,j}{}' |\mathbf{y}_i - \mathbf{y}_j|^{-3+\xi} \leq C < \infty \qquad \xi > 0 \tag{2.8}$$

$$N^{-3} \sum_{i,j,k}{}' |\mathbf{y}_i - \mathbf{y}_j|^{-2} |\mathbf{y}_j - \mathbf{y}_K|^{-2} \leq C < \infty\ . \tag{2.9}$$

$\left(\sum_{i,j}{}' = \sum_{\substack{i,j\\ j\neq i}} \right)$.

Then

$$\lim_{N\to\infty} P_{\Omega^N}(\Omega^N - \Omega_1^N) = 0.$$

Lemma 2.2

Let A be a bounded domain in R^3,

$$\begin{aligned} \mu\Delta \mathbf{z} - \lambda \mathbf{z} &= \nabla P \quad \text{in } A \\ \nabla\cdot\mathbf{z} &= 0 \quad \text{in } A \\ \mathbf{z}|_{\partial A} &= U. \end{aligned} \tag{2.10}$$

Set

$$B(u) = \mu \int_A u_{i,j} u_{i,j} + \lambda \int_A u_i\, u_i ,$$

and let G be the class of piecewise differentiable divergence free vector fields in A, satisfying $\mathbf{u}|_{\partial A} = U$. Then

$$B(\mathbf{z}) = \min_{\mathbf{u}\,\varepsilon\, G} B(\mathbf{u}).$$

Let $\psi(\mathbf{x},\mathbf{y})$ be Green's function of (2.8) in D, and $\tilde{\psi}$ the 3N×3N matrix which is composed of 3×3 blocks, the (i,j) block being $\psi(\mathbf{y}_i,\mathbf{y}_j)$ for $i\neq j$ and the 3×3 zero matrix when $i = j$. Then

Lemma 2.3

$$\| \frac{1}{N}\ \tilde{\psi}\ \| \leq C\lambda^{-\frac{b}{2}} \qquad \forall\ b < \frac{1}{2}$$

In the spirit of Theorem (2.1) we are going to show:

Theorem 2.2 $(\gamma > 0)$

Let F and ρ be as in Theorem (2.1). Then: for every $\varepsilon > 0$, there is N_0 s.t. for every $N \geq N_0$

$$P_{\Omega^N} \{\|\mathbf{u}^N - \mathbf{v}\|_{L_2} < \varepsilon\} > 1 - \varepsilon$$

for $\lambda \geq \lambda_0 N^{4\gamma}$ and $\gamma \in [0, \frac{1}{12})$.

Remarks

1) Theorems 2.1, 2.2 can be proved under weaker conditions. It suffices, for example, that u_0, $F \in L_q$, $q > 3$.

2) We first prove convergence on Ω_1^N and then apply Lemma 2.1.

3) The assumption that $\{y_j\}$ are fixed is not very realistic in R^3 but we still believe that our model is useful to the understanding of hydrodynamical interactions. Similar results can be formulated for domains in R^2 where such materials can be readily realized.

4) Throughout the paper we use ε and ζ to denote arbitrary small positive quantities, while C stands for a generic constant.

III. Proof of the Main Theorem

Our proof is based on the point interaction approximation which was first suggested by Ozawa [9,10] for diffusion problems, and then was used by Rubinstein [12] to solve for $\gamma = 0$. The specific formulation we use is due to Figari et. al. [3].

We introduce the following function as an approximation for u^N:

$$\eta^N = \int[\psi(x,y)F(x) - \frac{\beta}{N}\tilde{\Phi}^T(y)(\frac{\beta}{N}\tilde{\psi} + I_{3N})^{-1}\tilde{\Phi}(x)F(x)]dx, \qquad (3.1)$$

where

$$\tilde{\Phi} = \begin{matrix} \Psi(x,y_1) \\ \vdots \\ \Psi(x,y_N) \end{matrix},$$

and $\beta = 6\pi\mu\alpha$

η^N is the point sources approximation. We replace each sphere with a singularity which is proportional to Green's function of D, and the weights q_x^j are chosen such that the surface average of η^N is zero on each sphere.

Choosing $\lambda = \lambda_0 N^{(4+\xi)\gamma}$, $(\xi>0)$, and using Lemma 2.3 we see that

$$\|\frac{\beta}{N}\tilde{\psi}\| < 1 \text{ for } \lambda_0 \text{ large enough.}$$

Lemma 3.1

$$\|\eta^N - u^N\|_{L_2(D^N)} \leq CN^{-\frac{1}{6}+2\gamma} .$$

Proof

Let $\delta^N = \eta^N - u^N$. δ^N solves then

$$\mu\Delta\delta^N - \lambda\delta^N = \Delta P^N \qquad \text{in } D^N \tag{3.2}$$

$$\nabla\cdot\delta^N = 0$$

$$\delta^N(y) = \eta^N(y) \quad \text{for } y \in S_j^N \quad j = 1,2,,N .$$

Let $\{T_i\}$ $i = 1,2,,N$ be the following domains:

$$T_i = \{x;\ \frac{\alpha}{N} \leq |x-y_i| \leq \frac{2\alpha}{N}\} . \tag{3.3}$$

We assume that $\{y_i\}$ belong to Ω_i^N so that $\{T_i\}$ are security domains.

In each T_i we set τ^i to be the solution of

$$\nabla\cdot\tau^i = 0 \tag{3.4}$$

$$\tau^i|_{S_j^N} = \eta^N|_{S_j^N} ,\ \tau^i(y) = 0 \quad |y-y_i| > \frac{2\alpha}{N} ,$$

and define

$$\tau^N = \sum_{i=1}^N \tau^i .$$

From Lemma 2.2 we have the following chain of inequalities:

$$\lambda^{1/2}\|\delta^N\|_{L_2} \leq B^{1/2}(\delta^N) \leq B^{1/2}(\tau^N) \leq \lambda^{1/2}\|\tau^N\|_{H^1} , \tag{3.5}$$

hence

$$\|\delta^N\|_{L_2} \leq (\sum_{j=1}^N \|\tau^i\|_{H_1}^2)^{1/2}$$

In [12] we have shown how to estimate $\|\tau^i\|_{H^1}$ by first controlling η^N on S_i^N, Then "blowing up" T_i to a smooth domain with a size of $O(1)$, and finally applying standard a priori estimates. The idea behind the estimates for η^N is that since η^N has zero surface average on every sphere, it will be small everywhere there, unless there are large variations because two (or more) spheres happen to be close to each other. But this happens with a very small probability due to Lemma 2.1. Without spelling out the details we obtain

$$\|\delta^N\|_{L_2} \leq CN^{-\frac{1}{6}+2\gamma} \qquad \text{Q.E.D.}$$

Next, we show convergence of η^N to $\underset{\sim}{v}$: This is done by comparing the resolvent expansions. Expanding the integrand of (3.1) (recall Lemma 2.3), the general term is of the form

$$\int \left(\frac{\beta}{N}\right)^s \underset{\approx}{\Psi}_{ii_1}(y, y_{j_1}) \underset{\approx}{\Psi}_{i_1 i_2}(y_{j_1}, y_{j_2}) \cdots \underset{\approx}{\Psi}_{i_{s-1} i_s}(y_{j_{s-1}}, y_{j_s}) \cdot$$

$$\underset{\approx}{\Psi}_{i_s i_{s+1}}(y_{j_s}, z) F_{i_{s+1}}(z) dz .$$

Write

$$\eta^N = \nu^N + \zeta^N ,$$

where ν^N contains all the terms of η^N which have only distinct points $\{y_j\}$. For every $f \in C_0(D)$ we denote by $\eta^N(f)$ the solution of (2.4) with nonhomogeneous term f. (Note that we consider now η^N for $x \in D$). We want to estimate

$$N^{1-\sigma}(h, \zeta^N(f)) \quad \text{for} \quad h \in L_2(D),\ f \in C_0(D). \tag{3.7}$$

Let the first point which repeats itself be at the nth place at its first appearance, and then again at the $n_1 + n_2^{th}$ place. The contribution of such a term to (3.7) is

$$T_{n_1 n_2} = \frac{\beta^{n+1}}{N^{n-1+\sigma}} \sum_{\substack{i,j,p,k,\\ r,q,\ell=1}}^{3N} (\underset{\sim}{h\tilde{\Phi}})_i (\underset{\approx}{\tilde{\Psi}}^{n_1-1})_{ip} (\underset{\approx}{\tilde{\Psi}})_{pk} (\underset{\approx}{\tilde{\Psi}})_{k\ell} \cdot$$

$$(\underset{\approx}{\tilde{\Psi}})_{\ell q} (\underset{\approx}{\tilde{\Psi}})_{qk} (\underset{\approx}{\tilde{\Psi}})_{k'r} (\underset{\approx}{\tilde{\Psi}}^{n-n_1-n_2-1})_{rj} (\tilde{\Phi}_f)_j$$

where $|k-k'| < 3$.

Using standard potential estimates ([18]), Cauchy Schwartz inequality and Lemma 2.1 we find

$$T_{n_1 n_2} \leq C N^{5\gamma} \|h\|_2 \|f\|_2 (C\lambda_0)^{-nb/2} \cdot N^{-\sigma} .$$

Since there are at most n^2 such terms for a chain of length n, and for λ_0 large enough

$$\sum n^2 (c\lambda_0)^{-nb/2}$$

converges, we obtain

Lemma 3.2

$$\|\zeta^N\|_{L_2}(D) \leq C N^{-1+5\gamma} .$$

This was the crucial step in the averaging procedure, since now we can easily compute the expectation of $\mathbf{v}^N$, and for each term of the form (3.6) one finds:

$$N^{-S} \beta^S \int \underset{\approx}{\mathbf{T}}_{i i_1}(y,z_1)\rho(z_1)\underset{\approx}{\mathbf{T}}_{i_1 i_2}(z_1,z_2)\rho(z_2) \cdots \underset{\approx}{\mathbf{T}}_{i_s i_{s+1}}(z_s,z_{s+1}). \quad (3.8)$$

$$\mathbf{F}_{i_{s+1}}(z_{s+1})dz_1 \cdots dz_{s+1} .$$

which is precisely N^{-S} times the corresponding term in the expansion for $\mathbf{v}$. We only have to count how many terms like (3.6) appear to conclude (cf. [3] or [12]):

$$\lim_{N\to\infty} P_{\Omega N} \{N^{-\frac{1}{2}+\zeta} \|\mathbf{v}^N - \mathbf{v}\|_{L_2} < \varepsilon\} = 1. \quad (3.9)$$

Extending u^N to $D\text{-}D^N$ by defining it to be identically zero there (Observe that $\mathrm{Vol}(D\text{-}D^N) = O(N^{-23/12})$, and combining Lemma 3.1, Lemma 3.2 and (3.9) we get $(N > N_0(\varepsilon))$.

$$P_{\Omega^N} \{\|\mathbf{u}^N - \mathbf{v}\|_{L_2} < \varepsilon\} > 1-\varepsilon \quad (3.10)$$

Remark

The steady state case ($\lambda=0$) can be treated similarly: we add to both sides of (2.4) $-\lambda \underset{\sim}{u}^N$, and redefine the inhomogeneous term $\underset{\sim}{F}$. While the proof is unchanged in the Brinkman scaling ($\gamma=0$), we have to restrict $\gamma \in (0,1/60)$ in the Darcy scaling ($\gamma>0$).

IV. Spectral Interpretation

We give now some spectral interpretation of the previous results in the spirit of Ozawa ([9], [10]). Decompose $L_2 = J + J^{\perp}$ where

$$J = \{\mathbf{u} \in L_2(D);\ \nabla \cdot \mathbf{u} = 0,\ \mathbf{u}|_{\partial D} = 0\}$$

$$J^{\perp} = \{\mathbf{u} \in L_2(D);\ \mathbf{u} = \nabla P,\ P \in H^1(D)\},$$

and we denote by M_J the projection $M_J : L_2 \to J$. For every $\mathbf{f} \in L_2(D)$ we define an operator $\psi^{\lambda}(\mathbf{f})$ such that $\mathbf{u} = \psi^{\lambda}(\mathbf{f})$ solves

$$\begin{aligned} \Delta\mathbf{u} - \lambda\mathbf{u} &= \nabla P + \mathbf{f} \\ \nabla \cdot \mathbf{u} &= 0 \\ \mathbf{u}\,|_{\partial D} &= 0. \end{aligned} \tag{4.1}$$

Alternatively, one can set $\tilde{\Delta}_{\lambda} = M_J(\Delta - \lambda)$, and then

$$\tilde{\Delta}_{\lambda}^{-1} = \psi^{\lambda}.$$

It is well known that $-\psi^{\lambda}$ is a self-adjoint, compact and positive definite operator on L_2, and hence $-\Delta_{\lambda}$ has discrete eigenvalues

$$0 < \mu_1^{\lambda} < \mu_2^{\lambda} < \mu_3^{\lambda} \ldots,\ \mu_K^{\lambda} \to \infty \text{ as } K \to \infty, \tag{4.2}$$

and a complete set of eigenfunctions $\{\mathbf{g}_n^{\lambda}\}$. We want to analyse the spectrum of $\tilde{\Delta}_{N,\lambda}$ (i.e. $\tilde{\Delta}_{\lambda}$ restricted to D^N). For the sake of simplicity we limit the discussion to the case where $\gamma = 0$, ρ is constant, and the eigenvalues of $-\Delta + 6\pi\mu\alpha_0\rho$ are distinct. Set

$$b = b(\lambda) = \lambda + 6\pi\mu\alpha_0\rho,$$

then we have

Theorem 4.1

$$\lim_{N\to\infty} P_{\Omega^N} \{|\mu_i^{N,\lambda} - \mu_i^{b(\lambda)}| < \varepsilon\} = 1 \quad \forall\, \varepsilon > 0,\ \lambda \geq 0$$

Proof

Let χ_N be the characteristic function of D^N. Then

$$(g_i^{N,\lambda},\ \psi^{N,\lambda}\ \chi_N\ g_n^b) = -\ (\mu_i^{N,\lambda})^{-1}\ (g_i^{N,\lambda}\ ,\ g_n^b)$$

$$(g_i^{N,\lambda},\ \psi^b\ g_n^b) = -\ (\mu_n^b)^{-1}\ (g_i^{N,\lambda},\ g_n^b)\ ,$$

hence

$$(g_i^{N,\lambda},\ g_n^b) = \mu_i^{N,\lambda}\ \mu_n^b\ (\mu_n^b - \mu_i^{N,\lambda})^{-1}(g_i^{N,\lambda},(\psi^{N,\lambda}\chi_N - \psi^b)g_n^b)$$

From Theorem (2.1) (and considering $\{\mathbf{y}_j\} \in \Omega_1^N$),

$$|(g_i^{N,\lambda},\ g_n^b)| \leq |\mu_i^{N,\lambda} - \mu_n^b|^{-1}\ \mu_i^{N,\lambda}\ \mu_n^b\ \|g_n^b\|_q\ \varepsilon(N)$$

where $q > 3$ and $\varepsilon(N) \to 0$ as $N \to \infty$. But $g_n^b = -\ \mu_n^b\ \psi^b(g_n^b)$, so

$$\|g_n^b\|_\infty \leq C\mu_n^b\ ,\ \text{hence}$$

$$|(g_i^{N,\lambda},\ g_n^b)| \leq \varepsilon(N)\,|\mu_n^b - \mu_i^{N,\lambda}|^{-1}\ \mu_i^{N,\lambda}\ (\mu_n^b)^2\ ,$$

and either $\lim\limits_{N\to\infty} \mu_i^{N,\lambda} = \mu_n^b$ or $|(g_i^{N,\lambda},\ g_n^b)| \underset{N}{\to} 0.$

To complete the proof we note that $\{g_n^b\}$ and $\{g_i^{N,\lambda}\}$ are complete sets. Q.E.D.

Remarks

i). The presence of the obstacle caused a shift of $6\pi\mu\alpha_0$ in the spectrum. This is what we mean by "screening".

(ii) This shift in the spectrum is not a consequence of the randomness. Similar results hold for periodic structures.

V. Applications to Flow in Porous Media

Considering the geometry described at Section II as a model for porous media, equation (2.3) an be regarded as the effective (macroscopic) equations. We distinguish between 3 cases:

(i) $\gamma < 0$. The effective equation is again Stokes equation. The obstacles have a negligible effect on the flow.

(ii) $\gamma = 0$. The flow is described macroscopically by Brinkman's equation

$$\mu\Delta\mathbf{v} - \mu\underset{\approx}{K}^{-1}\,\mathbf{v} - \nabla P + \mathbf{F} = 0$$

$$\nabla\cdot\mathbf{v} = 0$$

$$\mathbf{v}|_{\partial D} = 0,$$

$$\underset{\sim}{K}^{-1} = 6\pi\alpha_0\rho(x)\,\underset{\sim}{I}\,. \tag{5.2}$$

(iii) $\gamma \in (0, \frac{1}{60})$. There it can be shown by an analysis similar to Section **IV** that the shift in the spectrum is $O(N^{\gamma})$. The macroscopic equation is:

$$\mu\Delta\mathbf{v} - N^{\gamma}\,6\pi\mu\alpha_0\,\rho(\mathbf{x})\mathbf{v} - \nabla P + \mathbf{F} = 0 \tag{5.2}$$

$$\nabla\cdot\mathbf{v} = 0$$

$$\mathbf{v}\,|_{\partial D} = 0.$$

This is a singular perturbation problem, and in order to study it we introduce the spaces:

$$Q_1 = \{\mathbf{v} \in H^1(D), \quad \nabla\cdot\mathbf{v} = 0\}$$

$$Q_2 = \{\mathbf{v} \in L_2(D)\,,\ \nabla\cdot\mathbf{v} = 0,\ \mathbf{v}\cdot n|_{\partial D} = 0\}$$

While (5.2) is considered for $\mathbf{v}$ in Q_1, we let $\mathbf{U}(\mathbf{x}) \in Q_2$ be the solution of

$$-\,6\,\pi\mu\alpha_0\rho(\mathbf{x})\mathbf{U} = \nabla P - \mathbf{F} \tag{5.3}$$

$$\mathbf{U}\cdot\hat{n}|_{\partial D} = 0\,.$$

Using standard methods of singular perturbation ([8]) we can prove

$$\lim_{N\to\infty} \| N^{\gamma} \mathbf{v} - \mathbf{U} \|_{Q_2} = 0.$$

Finally we see that under the conditions of Theorem 2.2

$$\lim_{N\to\infty} P_{\Omega^N} \{ \| N^{\gamma} \mathbf{u}^N(\mathbf{x}) - \mathbf{U}(\mathbf{x}) \|_{Q_2} < \varepsilon \} = 1.$$

$$\forall\ \varepsilon > 0, \quad \text{and} \quad \gamma \in (0,\tfrac{1}{60}).$$

This means that we can expand

$$\mathbf{u}^N = N^{-\gamma}\, \mathbf{u}^{N,0} + N^{-2\gamma}\, \mathbf{u}^{N,1} + \ldots$$

and $\mathbf{u}^{N,0}$ converges strongly (Q_2) to the solution of the <u>Darcy</u> <u>equation</u> (5.3)

VI. Discussion

We make here some concluding remarks:

1) The fact that Darcy law is a singular perturbation limit of Brinkman's equation was realised by many people (e.g. [14]). What we have shown here, however, is the full transition, in a random setup, from Stokes equation on the microscopic level to Brinkman-Darcy type equation as the macroscopic laws.

2) The point interaction approximation was stretched here to its limit, and even this extension applies only to dilute systems. We are currently studying dense configurations by other methods.

Acknowledgment

I am grateful to George Papanicolaou and Andrea Prosperetti for their valuable suggestions. This research was supported in part by the IMA with funds provided by the NSF and the ARO.

References

1. H.C. Brinkman, Appl. Sci. Res. A1, 27. (1947).

2. S. Childress, J. Chem. Phys. 36, 2527. (1972)

3. R. Figari, E. Orlandi and S. Teta, J. Stat. Phys. 41, 465. (1985).

4. K.F. Freed and M. Muthukumar, J. Chem. Phys. 68, 2088. (1978).

5. J. Happel and H. Brenner, Low Reynolds Number Hydrodynamics, Prentice-Hall (1965).

6. E.J. Hinch, J. Fluid. Mech. 83, 695. (1977).

7. O.A. Ladyzhenskaya, The Mathematical theory of Viscous Incompressible Flow, Gordon and Breach (1969).

8. J.L. Lions, Perturbations Singuliéres dans les Problémes aux Limites et en Controle Optimal. Lecture Notes in Mathematics Vol. 323, Springer-Verlag (1973).

9. S. Ozawa, Comm. Math. Phys. 91, 473 (1983).

10. S. Ozawa, Comm. Math. Phys. 94, 421 (1984).

11. G.C. Papanicolaou, in: Les Méthods de L'homogénéisation: Theorie et Application en Physics, INRIA (1985).

12. J. Rubinstein, J. Stat. Phys. 44, 849. (1986)

13. J. Rubinstein, J. Fluid Mech. 170, 379. (1986)

14. E. Sanchez-Palencia, Int. J. Eng. Sci. 20, 1291. (1982).

15. C.K.W. Tam, J. Fluid Mech. 38, 537. (1969).

16. R. Temam, Navier Stokes Equations. North Holland (1977).

17. R.A. Adams, Sobolev Spaces, Academic Press (1975).

18. D. Gilbarg and N.S. Trudinger, Elliptic Partial Differential Equations of Second Order, Springer-Verlag (1983).

INTERACTING BROWNIAN PARTICLES : A STUDY OF DYSON'S MODEL

Herbert Spohn

Theoretische Physik, Universität München
Theresienstr. 37, D - 8000 München 2 , Germany

Abstract. We study the equilibrium fluctuations of Brownian particles in one dimension interacting through the pair force $1/x$. In the hydrodynamic limit the structure function is $S(k,t) = (|k|/2\pi)\exp[-|t||k|\pi\rho]$ with ρ the density of particles and the fluctuation field is Gaussian.

1. Dyson's Model

We consider Brownian particles in one dimension interacting through the pair force $1/x$. The equations of motion are

$$dx_j(t) = \sum_{i,i\neq j} \frac{1}{x_j(t) - x_i(t)} dt + db_j(t) \tag{1.1}$$

with label $j = 0,\pm 1,\ldots$. $x_j(t) \in R$ is the position of the j-th particle at time t. $\{b_j(t), j \in \mathbb{Z}\}$ are a collection of independent standard Brownian motions. The pair force between particles is repulsive. Needless to say that the force on the j-th particle is rather singular in fact both for short and large distances. The small x singularity implies that particles cannot cross each other (with probability one) and therefore their order is preserved in time. Because of the large x singularity the force on the j-th particle is not summable. Therefore the sum should be thought of as an improper sum, in the sense that the summation is restricted to those i's such that $|x_i(t) - x_j(t)| < \kappa$ and subsequently $\kappa \to \infty$. This shows that the dynamics can exist only for initial configurations having the same asymptotic density to the right and left. We will prove the existence of the equilibrium dynamics, in sense of the appropriate Markov semigroup.

Our goal is to understand the time-dependent equilibrium fluctuations of the system on a large space-time scale (hydrodynamic limit). The standard fluctuation theory is based on the assumption of short range forces, cf. Section 2. Certainly $1/x$ does not qualify and the standard theory is expected to fail. The problem is how and what is then the correct theory.

Our analysis exploits very special properties of the model defined by (1.1). In fact, we cannot even allow to replace the force $1/x$ by $\beta/2x$, $\beta > 0$, with $\beta \neq 2$.

I will call (1.1) Dyson's model. In the early '60 Dyson studied random matrices connected to the statistics of nuclear energy levels [1]. He made the following amusing observation [2]: Let $M(t)$ be a complex, symmetric NxN matrix. We assume that the matrix elements $\{M_{ij}(t), i,j = 1,\ldots,N\}$ are random. The diagonal elements, which are real, are governed by the Ornstein-Uhlenbeck process

$$dM_{jj}(t) = -\lambda^2 M_{jj}(t)\,dt + \sqrt{2}\,dW_{jj}(t) \tag{1.2}$$

with $W_{jj}(t)$ independent Brownian motions. The real and imaginary parts of the off-diagonal matrix elements, say in the lower triangle, are independent and governed by

$$dM^{\sigma}_{ij}(t) = -\lambda^2 M^{\sigma}_{ij}(t)\,dt + dW^{\sigma}_{ij}(t) \quad , \tag{1.3}$$

σ=real, imaginary, with $W^{\sigma}_{ij}(t)$ independent Brownian motions. Since $M(t)$ is symmetric, it has N real eigenvalues $x_1(t) \leq \ldots \leq x_N(t)$. Dyson shows, in law, the eigenvalues satisfy

$$dx_j(t) = -\lambda^2 x_j(t)\,dt + \sum_{i=1,i\neq j}^{N} \frac{1}{x_j(t) - x_i(t)}\,dt + db_j(t) \quad . \tag{1.4}$$

With probability one the eigenvalues do not cross and therefore the ordering is preserved in the course of time.

The proof is actually not difficult. Writing the stochastic differential for the eigenvalues $x_j(t)$ we have to know their change to first and second order under a small change of the matrix $M(t)$. This is standard perturbation theory for matrices (assuming non-degeneracy). The second order change results in the interaction. The first order change gives the first term and the noise. One verifies that the noise has the statistics of independent Brownian motions.

(1.4) is the motion of N interacting Brownian particles in a common harmonic external potential $\frac{1}{2}\lambda^2 x^2$. We are interested here in the limit where the strength of the harmonic potential tends to zero, but at the same time the number of particles is increased such that the average density remains constant. This is ensured by

$$\lambda \to 0,\ N \to \infty\ ,\ \lambda N^2 = 2(\pi\rho)^3/3 \quad , \tag{1.5}$$

where ρ is the average density of particles. In terms of the random matrices the limit is taken in such a way that, typically, the spectrum of $M(t)$ becomes unbounded to both sides and remains purely discrete and nondegenerate.

On a technical level I do not use the matrix representation of the dynamics (1.4).

To give a brief outline : In Section 2 we remind the reader of the standard fluctuation theory for short range forces and indicate its generalization. Section 3 deals with the equilibrium measure for (1.4) in the limit (1.5). In Section 4 we establish the existence of the equilibrium dynamics (actually in a rather weak sense). Sections 5 and 6 study the scaling limit. In the final section we point out the connection to interface models of equilibrium statistical mechanics.

2. Brownian Particles With Short Range Forces

As in (1.1) the dynamics of interacting Brownian particles is given by

$$dx_j(t) = -\frac{1}{2}\sum_{i,i\neq j} V'(x_j(t) - x_i(t))dt + db_j(t) \quad . \tag{2.1}$$

The only difference to (1.1) is a pair force $-\frac{1}{2}V'(x)$ where the potential V is a smooth function of compact support, $V(x)=V(-x)$. Let me further assume $V\geq 0$, $V(0)>0$. (For sake of comparison we restrict ourselves to one dimension, although the results to be mentioned hold for arbitrary dimension. Also the assumptions on V could be weakened.) For given fugacity z, $z > 0$, there is then a unique Gibbs measure for the potential V. This Gibbs measure is denoted by μ_z. To z there corresponds a unique average density $\rho = \rho(z)$. For μ_z - almost all initial conditions the solution to (2.1) exists [3]. μ_z is invariant under the dynamics. In that sense the equilibrium dynamics is well-defined. We consider here only the stationary process, i.e. (2.1) with initial data distributed according to μ_z.

To study the equilibrium fluctuations on a large space-time scale it is convenient to introduce the fluctuation field

$$\xi^\varepsilon(f,t) = \varepsilon^{1/2}\Big(\sum_j f(\varepsilon x_j(t)) - \rho\int dx f(\varepsilon x)\Big) \quad . \tag{2.2}$$

Here f is a rapidly decreasing test function. $\xi^\varepsilon(\cdot,\cdot)$ is a generalized process stationary in space-time. To make the following a theorem [4] we require the density to be small enough, in the sense

$$z \leqq 0.1/\int dx(1 - e^{-V(x)}). \tag{2.3}$$

Then $\xi^{\varepsilon}(f,t)$ converges in law to a limit process,

$$\lim_{\varepsilon\to 0} \xi^{\varepsilon}(f,t) = \xi(f,t) \quad , \tag{2.4}$$

and $\xi(f,t)$ is a Gaussian process with mean zero and covariance

$$\langle\xi(f,t)\xi(h,s)\rangle = \chi\int dk\hat{f}(k)\hat{h}(-k)\exp[-k^2|t-s|\rho/2\chi] \quad . \tag{2.5}$$

χ is the compressibility for μ_z. $\hat{}$ denotes Fourier transform.

χ is merely a normalization to make (2.4), (2.5) correct for t=0. The dynamical content of (2.4), (2.5) has a simple physical interpretation : If we impose on the equilibrium system at $t = 0$ a fluctuation localized near the origin, then this fluctuation will spread out according to the fundamental solution of the diffusion equation. However the speed of spreading, ρ/χ, depends in a complicated way on the interaction between particles. Only if $\rho \to 0$, implying $\rho/\chi \to 1$, we recover the spreading for independent Brownian particles.

The limit Gaussian process can also be thought of as a stationary infinite-dimensional Ornstein-Uhlenbeck process governed by

$$d\xi(x,t) = \frac{\rho}{2\chi}\frac{\partial^2}{\partial x^2}\xi(x,t)dt + \sqrt{\rho}\,\frac{\partial}{\partial x}dW(x,t) \quad , \tag{2.6}$$

where $dW(x,t)$ is normalized white noise in space-time.

How should (2.4), (2.5) be generalized to long range forces ? To explain the physical heuristics we have to introduce the dynamic structure function defined by

$$\int dk|\hat{f}(k)|^2 S(k,t) = \langle(\sum_j f(x_j(t)))(\sum_i f(x_i(0)))\rangle - (\rho\int dx f(x))^2, \tag{2.7}$$

expectation in the stationary process. In general, $S(k,t)dk$ is a measure. Let me assume here that it has a density jointly continuous in k,t. The static structure function of the Gibbs measure μ_z is $S(k) = S(k,t=0)$. Then, by definition

$$\chi = S(0) \quad . \tag{2.8}$$

More precisely now, the problem is to generalize (2.4), (2.5) in case either $\chi = 0$ or $\chi = \infty$ and therefore the naive application of (2.5) makes no sense. There is a physical theory, so-called mode-coupling theory, which covers also cases with χ either zero or infinite. For diffusive systems, under consideration here, the claim is that for small k and large t

$$S(k,t) \cong S(k)\exp[-k^2|t|\rho/2S(k)] \quad . \tag{2.9}$$

The k^2 reflects the diffusive spreading and S(k) is its static modification.

Now if

$$\lim_{k\to 0} S(k) = S(0) = \chi > 0 \tag{2.10}$$

exists, then we are back to the case discussed before, namely

$$\lim_{\varepsilon\to 0} S(\varepsilon k,\varepsilon^{-2}t) = \chi\exp[-k^2|t|\rho/2\chi] \quad , \tag{2.11}$$

which rephrases the convergence of second moments in (2.4), (2.5). If the system has strong static fluctuations, then S(k) diverges for small k indicating criticality. Of course the small k divergence modifies the scaling behavior of S(k,t). Two examples, confirming the recipe (2.9), have been studied on a mathematical level. The first one is diffusive dynamics of harmonic and anharmonic crystals for dimension $d \geq 3$. For these models $S(k) \cong c/k^2$ for small k, a highly nontrivial result for the anharmonic crystal [5,6]. Following (2.9) therefore

$$\lim_{\varepsilon\to 0} \varepsilon^2 S(\varepsilon k,\varepsilon^{-4}t) = (c/k^2)\exp[-k^4|t|/c] \quad . \tag{2.12}$$

($\rho=1$ for this case.) This is a theorem [7] for couplings of the form $\sum_x((-\Delta)^\alpha\phi_x)^4$, $\alpha>1/2$, cf. [8] for the static case. For couplings $\sum_x(\nabla_x\phi)^4$, i.e. $\alpha=1/2$, a proof of the dynamic scaling remains open. The second one is the voter model in dimension $d\geq 3$. The voter model has no conservation law. Therefore (2.9) should be modified to $S(k,t) \cong S(k)\exp[-|t|/S(k)]$. Since $S(k) \cong c/k^2$, for small k, we find

$$\lim_{\varepsilon\to 0} \varepsilon^2 S(\varepsilon k,\varepsilon^{-2}t) = (c/k^2)\exp[-|t|k^2/c] \tag{2.13}$$

in accordance with the theorems of [9,10]. Note the modified scaling of S(k,t) compared to (2.11). If the fluctuation fields are normalized in accordance, they converge again to an infinite dimensional Ornstein-Uhlenbeck process.

As we will see, for Dyson's model something completely different happens. Due to the strong, long-ranged repulsion between particles the large scale fluctuations are heavily suppressed and

$$\lim_{k\to 0} S(k) = 0 \quad . \tag{2.14}$$

In fact for small k

$$S(k) = |k|/2\pi \quad . \tag{2.15}$$

If we believe in the heuristic recipe (2.9), we would conclude

$$\lim_{\varepsilon\to 0} \varepsilon^{-1} S(\varepsilon k, \varepsilon^{-1} t) = (|k|/2\pi)\exp[-|k||t|\rho\pi] \tag{2.16}$$

for Dyson's model. We will prove that this is in fact correct.

I find it rather surprising that a simple prescription as (2.9) apparently covers such a wide range of models.

3. The Infinite Volume Equilibrium Measure

The equilibrium measure of (1.4) is

$$\frac{1}{Z} \prod_{1\leq i<j\leq N} (x_i - x_j)^2 \prod_{j=1}^{N} \exp[-\lambda^2 x_j^2] dx_j = \mu_N(x_1,\ldots,x_N) \prod_{j=1}^{N} dx_j \quad . \tag{3.1}$$

Z normalizes the measure to one. According to (1.5) we fix the density $\rho>0$ once for all and set $\lambda = 2(\pi\rho)^3/3N^2$. Our notation suppresses the dependence on ρ. Averages with respect to μ_N are denoted by $\langle\cdot\rangle_N$. The properties of μ_N are rather well understood [1].

We want to prove that the infinite volume limit, $N\to\infty$, of μ_N exists and defines a measure on the space of all locally finite configurations [11], denoted here by X . It is convenient to define a particular class of (unbounded) functions on X . Let $\underline{x} \in X$ and define

$$n(f)(\underline{x}) = \sum_j f(x_j) \quad , \tag{3.2}$$

where f is a smooth test function of compact support. We denote by P the algebra of finite polynomials of the form (3.2). Let

$$\rho_n^{(N)}(x_1,\ldots,x_n) = \frac{N!}{(N-n)!} \int dx_{n+1}\ldots dx_N \mu_N(x_1,\ldots,x_N) \tag{3.3}$$

be the n-th correlation function of μ_N. Note that

$$\rho_n^{(N)}(y_1,\ldots,y_n) = \langle \sum_{j_1 \neq \ldots \neq j_n} \prod_{i=1}^{n} \delta(x_{j_i} - y_i) \rangle_N \qquad (3.4)$$

and therefore $\langle \prod_{j=1}^{n} n(f_j) \rangle_N$ is simply related to the first n correlation functions. Finally we define

$$R(x - y) = \frac{1}{2\pi} \int_{|k| \leq \pi\rho} dk \quad e^{ik(x-y)} \quad . \qquad (3.5)$$

Proposition 1. The correlation functions of μ_N have pointwise a limit

$$\lim_{N\to\infty} \rho_n^{(N)}(x_1,\ldots,x_n) = \rho_n(x_1,\ldots,x_n) \quad , \qquad (3.6)$$

where

$$\rho_n(x_1,\ldots,x_n) = \det R(x_i - x_j)\Big|_{i,j=1,\ldots,n} \quad . \qquad (3.7)$$

The correlation functions $\{\rho_n | n=1,2,\ldots\}$ determine a unique probability measure μ on X .

Proof: By [1] the finite N correlation functions are given by

$$\rho_n^{(N)}(x_1,\ldots,x_n) = \det R_N(x_i,x_j)\Big|_{i,j=1,\ldots,n} \qquad (3.8)$$

where

$$R_N(x,y) = \sum_{j=0}^{N-1} \psi_j(x)\psi_j(y) \qquad (3.9)$$

and $\psi_j, j = 0,1,\ldots,$ are the normalized eigenfunctions of the harmonic oscillator $-(d^2/2dx^2) + \frac{1}{2}\lambda^2 x^2$. $R_N \to R$ pointwise as $N\to\infty$ [1].
We have the bound

$$|\rho_n(x_1,\ldots,x_n)| \leq n!\rho^n \quad . \qquad (3.10)$$

By [12] the ρ_n's therefore determine uniquely a measure on X ∎

Expectations with respect to μ are denoted by $\langle\cdot\rangle$.

Let me define the Hilbert space $H = L^2(X, \mu)$. P is a dense subset of H.

The determinantal structure of the correlation functions suggests a fermionic structure. This is indeed the key to the analysis of Dyson's model. We therefore introduce the CAR algebra, A, over $L^2(R,dx)$ [13] with smeared creation and annihilation operators $a^+(f)=\int dx f(x)^* a^+(x)$, $a(f)=\int dx f(x) a(x)$ satisfying the canonical anticommutation relations

$$[a(f),a(h)]_+ = 0 \quad , \quad [a^+(f),a^+(h)]_+ = 0$$
$$[a(f),a^+(h)]_+ = 1\langle f|h\rangle \quad . \tag{3.11}$$

Here $[A,B]_+ = AB + BA$, $\langle\cdot|\cdot\rangle$ is the inner product in $L^2(R,dx)$ and 1 is the unit element of A. Readers not familiar with these notions should think of this simply as a convenient algebraic bookkeeping device with no further meaning attached to it. Let ω be the quasifree state on A defined by

$$\omega(a^+(f_1)\dots a^+(f_n)a(h_n)\dots a(h_1)) = \det\langle h_i|Rf_j\rangle\Big|_{i,j=1,\dots,n} \quad . \tag{3.12}$$

R is the linear operator with kernel (3.5). ω is the ground state of the infinitely extended ideal Fermi gas in one dimension with density ρ and Hamiltonian

$$H^F = -\int dx a^+(x)\frac{1}{2}\frac{d^2}{dx^2}a(x) \quad . \tag{3.13}$$

We define

$$n(f) = \int dx f(x) a^+(x) a(x) \tag{3.14}$$

and P the algebra of finite polynomials of n(f)'s. By (3.11) P is a commutative algebra. n(f) counts the number of fermions in the "region" defined by f. (The n(f)'s are unbounded and therefore not elements of A. Done properly, we should work in the GNS-representation of ω [13]. This is such a well understood chapter that we feel free to suppress it from our notation.)

Proposition 2. For any $\phi \in P$

$$\langle\phi\rangle = \omega(\phi) \quad . \tag{3.15}$$

Proof : Let $\phi = \prod_{j=1}^{n} n(f_j)$. To express $\langle\phi\rangle$ through correlation functions we divide the summation into $\{j_1 \neq \ldots \neq j_n\}$, one coinciding pair, etc. . This corresponds precisely to normal ordering $\Pi n(f_j)$, i.e. moving with the help of (3.11) all a^+'s to the left and a's to the right ∎

Remark . If in (3.7) we choose $R(x-y) = \int dk \nu(k) e^{ik(x-y)}$ with $0 \leq \nu(k) \leq 1$, then the so defined correlation functions still determine a unique probability measure on X , which in fact is translation invariant. The reason is that by (3.12) R defines a quasifree state ω_R on the CAR algebra A . Since we consider ω_R restricted to the commutative subalgebra P , the ρ_n's have to be the moments of some measure. I am not aware of a use of these measures outside of quantum mechanics.

The ground state of an ideal Fermi gas can be obtained also through other finite volume approximations than the one of Proposition 1, which corresponds to fermions in a weakening harmonic potential. The one which will be used later on is to put N fermions in a periodic box $[-L,L]$ with $N = 2L\rho$. Its ground state wave function squared is

$$\frac{1}{Z} \prod_{1 \leq i < j \leq N} \left(\sin\left(\frac{\pi}{2L}(x_i - x_j)\right)\right)^2 \prod_{j=1}^{N} \chi_{[-L,L]}(x_j) dx_j \quad . \tag{3.16}$$

This is the invariant measure of N Brownian particles on a ring $[-L,L]$ interacting through the pair force $\frac{\pi}{2L}\cot\frac{\pi}{2L}x$.

Proposition 3. The correlation functions of the measure (3.16) converge in the limit $N \to \infty$, $L = N/2\rho$, pointwise to (3.7).

4. The Infinite Volume Dynamics

Ideally we would like to show that (1.1) has a solution for μ - almost all initial conditions. This I do not know how to accomplish. A more modest approach would be to consider finite system correlations, e.g.

$$\langle n_t(f) n(f) \rangle_N \tag{4.1}$$

with $n_t(f) = \sum_{j=1}^{N} f(x_j(t))$, average in the stationary process (1.4), and to prove that they have a limit as $N \to \infty$. Again I do not know how to accomplish this. Instead, exploiting the fermionic structure, we essentially guess the definition of a strongly continuous semigroup T_t on $\mathcal{H} = L^2(X, \mu)$. Through its construction this semigroup has all the desired properties : It is selfadjoint, contracting, preserves positivity and mass (i.e. $T_t 1 = 1$). There exists then a generator, L, such that

$$T_t = e^{Lt} \quad . \tag{4.2}$$

We would like to show that for $\phi \in \mathcal{P}$

$$L\phi = \Big(\frac{1}{2}\sum_j \frac{d^2}{dx_j^2} + \sum_{i \neq j} \frac{1}{x_j - x_i}\frac{\partial}{\partial x_j}\Big)\phi \tag{4.3}$$

in $\mathcal{H}$.

Because $1/x$ is not integrable for large x, (4.3) is only formal. As an example let us consider $\phi = n(f)$. Then the symmetrized second term of (4.3) reads

$$\frac{1}{2}\sum_{i \neq j} \frac{1}{x_i - x_j}(f'(x_j) - f'(x_i)) \equiv \phi_\infty \quad . \tag{4.4}$$

For x_i close to x_j the summand is bounded. To define ϕ_∞ as an element of $\mathcal{H}$ we first have to introduce a cut-off as

$$\phi_\kappa = \frac{1}{2}\sum_{i \neq j} \frac{1}{x_j - x_i}\chi_\kappa(x_j - x_i)(f'(x_j) - f'(x_i)) \quad , \tag{4.5}$$

where χ_κ is the indicator function of the set $[-\kappa, \kappa]$. Then ϕ_κ is a Cauchy sequence whose limit would properly be called ϕ_∞. (For a non-symmetric cut-off a different limit would be obtained.) I have not carried through the proper technical construction and thus have to be satisfied with a somewhat weaker result.

<u>Proposition 4</u>. There exists a self-adjoint, positivity and mass preserving contraction semigroup $T_t, t \geq 0$, on $\mathcal{H}$ such that for any $\phi \in \mathcal{P}$

$$\lim_{t \to 0} \frac{1}{t}(\langle \phi T_t \phi \rangle - \langle \phi^2 \rangle) = -\frac{1}{2}\langle \sum_j \frac{\partial}{\partial x_j}\phi \frac{\partial}{\partial x_j}\phi \rangle \quad . \tag{4.6}$$

<u>Remark</u>. One may be inclined to define L through the Friedrichs extension of the Dirichlet form (4.6). I do not know how to prove that the corresponding semigroup preserves mass. The whole point here is that $\langle \phi T_t \phi \rangle$ for $\phi \in \mathcal{P}$ is in a way constructed explicitely.

Proof : (1) The δ-Bose gas. We consider N quantum mechanical particles on the circle [-L,L] interacting through an infinitely strong repulsive δ-potential. The Schrödinger operator of this system is given by

$$H_N = -\frac{1}{2}\sum_{j=1}^{N}\frac{d^2}{dx_j^2} \tag{4.7}$$

on $L^2([-L,L]^N)$ with periodic boundary conditions at $\pm L$ and Dirichlet boundary conditions whenever two particles touch, i.e.

$$\psi = 0\ , \quad \text{whenever} \quad x_i = x_j\ , \ i \neq j = 1,\dots,N\ . \tag{4.8}$$

We observe that the configuration space separates into (N-1)! disjoint regions, denoted by Δ_α. Let $\{p(j), j=1,\dots,N\}$ be a permutation of $\{1,\dots,N\}$. Two points are in the same Δ_α if they satisfy $-L \leq x_{p(1)} < \dots < x_{p(N)} < L$ up to a cyclic shift of the indices (e.g. $\{1,\dots,N\}$ to $\{N,1,2,\dots,N-1\}$). To Δ_α we associate sign(Δ_α), the sign of the corresponding permutation.

In each Δ_α the complete set of eigenfuctions for H_N can be obtained. Let ϕ_j be the eigenfunctions of $-(d^2/2dx^2)$ on [-L,L] with periodic boundary conditions, j = 1,2,... . Let $(x_1,\dots,x_n) \in \Delta_\alpha$. The eigenfunctions of H_N in Δ_α are labeled by subsets of $\{1,2,\dots\}$ containing precisely N elements. We denote them by $\{q\}$. Then the eigenfunctions are

$$\psi_{\{q\}}(x_1,\dots,x_N) = Z^{-1/2}\,\text{sign}(\Delta_\alpha)\,\det \sum_{\{q\}} \phi_q(x_j)^*\phi_q(x_i)\Big|_{i,j=1,\dots,N}\ . \tag{4.9}$$

with $Z^{-1/2}$ as normalization.

Let ψ_o be the ground state of the system. It corresponds to $\{q\} = \{1,\dots,N\}$. Let E_o be its energy. ψ_o^2 is given by (3.16). We denote expectations with respect to ψ_o^2 by $\langle\cdot\rangle_N$. Let F and G be bounded functions on $[-L,L]^N$ symmetric in their arguments. Then we define

$$\langle FT_t^{(N)}G\rangle_N = \langle\psi_o|Fe^{-t(H_N-E_o)}G\psi_o\rangle\ . \tag{4.10}$$

As a standard fact about Schrödinger operators, $T_t^{(N)}$ is a contraction semigroup which preserves positivity and mass.

(2) Fermionic representation. The $\psi_{\{q\}}$, with sign(Δ_α) omitted and considered as defined on all of $[-L,L]^N$, (which changes the normalization), are the eigenfunctions of an ideal Fermi gas with N particles on a circle [-L,L]. We expand (4.10) in a complete set of eigenfunctions as

$$<\psi_o|Fe^{-t(H_N-E_o)}G\psi_o> = \sum_{\{q\}} \exp[-t(E(\{q\})-E_o)]$$
$$\sum_\alpha \int_{\Delta_\alpha} d\underline{x}\psi_o(\underline{x})^* F(\underline{x})\psi_{\{q\}}(\underline{x}) \int_{\Delta_\alpha} d\underline{x}\psi_o(\underline{x})G(\underline{x})\psi_{\{q\}}(\underline{x})^* \quad . \tag{4.11}$$

We note that $\int_{\Delta_\alpha}$ ... are independent of sign(Δ_α) and take the same value in every ele- ment of the partition. Therefore the right hand side of (4.11) may be reexpressed as the corresponding expectation in the free Fermi gas.

Let ω_N be the ground state of the Fermi gas with N particles on the circle [-L,L]. Let

$$H_L^F = -\int_{-L}^{L} dxa^+(x)\frac{1}{2}\frac{d^2}{dx^2}a(x) \tag{4.12}$$

with periodic boundary conditions. Let us choose for F,G some elements $\tilde{\phi},\phi \in P$. Of course the corresponding test functions must be supported inside [-L,L]. Then, using (4.10),

$$<\tilde{\phi}T_t^{(N)}\phi>_N = \omega_N(\tilde{\phi}e^{-t(H_L^F-E_o)}\phi) \quad . \tag{4.13}$$

The limit $N\to\infty$ of the right hand side is well understood. It defines then the semigroup T_t of the Proposition. T_t inherits its properties from $T_t^{(N)}$.

Let me be slightly more explicit as to the definition of T_t. The key fact is that, since the Hamiltonian (4.12) is quadratic in a^+,a,

$$\exp[it\ H_L^F]a(f)\exp[-itH_L^F] = a(e^{-it\Delta/2}f) \tag{4.14}$$

with $\Delta=-d^2/dx^2$. Therefore, together with the rule (3.12), all time-dependent moments are given "explicitely". In particular, the infinite volume limit is controlled by the limit of $<f|Re^{-t\Delta/2}h>$ and similar terms. We apply (4.14) to the imaginary time ground state dynamics (4.13). Let $\phi,\tilde{\phi}$ be monomials. Then, $t\geq 0$,

$$\omega_N(\tilde{\phi}\exp[-t(H_L^F-E_o)]\phi)$$
$$= \omega_N(\prod_{j=1}^{m} n(f_j)\exp[-tH_L^F]\prod_{j=m+1}^{n} n(f_j)\exp[tH_L^F]) \quad . \tag{4.15}$$

At the right hand side we use (4.14). The expectations are bounded for $t\geq 0$. The convergence is best controlled in Fourier space. We choose the f_j's such that the $\hat{f}_j$'s have compact support. Then the limit $L\to\infty$

is merely the convergence of Fourier sums to Fourier integrals.

We define, in Fourier space with $\hat{f}$ of compact support,

$$n_t(f) = \int dk dk' \hat{f}(k-k') \exp[-t(k^2-k'^2)] a^+(k) a(k') \quad . \tag{4.16}$$

Here

$$a(k) = (2\pi)^{-1/2} \int dx e^{ikx} a(x) \quad , \quad a^+(k) = a(k)^* \quad . \tag{4.17}$$

The $\{a^+(k), a(k)\}$ satisfy also the rules of the CAR algebra. (4.16) should be considered only as a formal definition to be used only in conjunction with ground state expectation. We extend this definition to all of P . Then on P T_t is defined by

$$\langle \tilde{\phi} T_t \phi \rangle = \omega(\tilde{\phi} \phi_t) \quad . \tag{4.18}$$

(3) The Dirichlet form. By construction (4.6) holds on the circle $[-L,L]$ with periodic boundary conditions. For any $\phi \in P$

$$\lim_{N\to\infty} \omega_N\left(\sum_j \frac{\partial}{\partial x_j}\phi \frac{\partial}{\partial x_j}\phi\right) = \omega\left(\sum_j \frac{\partial}{\partial x_j}\phi \frac{\partial}{\partial x_j}\phi\right) \quad . \tag{4.19}$$

Because the ideal Fermi gas is quasifree the infinite volume limit is a consequence of the convergence of the two-point function. For the two-point function d/dt and $N\to\infty$ are interchangable. (I tried to check (4.6) directly from the definition. The resulting combinatorics turned out to be not very penetrable.) ■

(4.18), perhaps more explicitely (4.15), expresses a somewhat mysterious fact of Dyson's model. The function $T_t n(f)$ on X is certainly not in P . Still $n_t(f)$ is quadratic in the Fermi operators a, a^+. Therefore on the CAR algebra mononomials of a given degree remain so in the course of time.

5. Scaling Limit : Covariance

After these preparations it is easy to obtain $S(k,t)$ defined by (2.7). The static structure factor is

$$S(k) = \begin{cases} |k|/2\pi & \text{for} \quad |k| \leqq 2\pi\rho \\ \rho & \text{for} \quad |k| \geqq 2\pi\rho \quad . \end{cases} \tag{5.1}$$

In real space this translates into

$$\langle n(x)n(0)\rangle - \rho^2 = \rho\delta(x) - \frac{1-\cos 2\pi x}{2\pi x^2} \quad . \tag{5.2}$$

The first term is just the diagonal of the double sum. The second term shows that, because of the strong repulsion, particles are negatively correlated. There is a decay as $-1/x^2$, but superimposed with a lattice structure of lattice spacing $1/\rho$.

The time-dependent structure function is given as, cf.[14],

$$S(k,t) = \frac{1}{4\pi|k||t|}(\exp[-\omega_2|t|] - \exp[-\omega_1|t|]) \tag{5.3}$$

with

$$\omega_1 = (k^2 + 2\pi\rho|k|)/2 \ , \quad \omega_2 = |k^2 - 2\pi\rho|k||/2 \quad . \tag{5.4}$$

Clearly, the scaling for small k and large t is

$$\lim_{\varepsilon\to 0} \varepsilon^{-1}S(\varepsilon k,\varepsilon^{-1}t) = \frac{|k|}{2\pi}\, e^{-|k||t|\pi\rho} \quad . \tag{5.5}$$

This is in accordance with the rule (2.9). In space-time (5.5) reads

$$[(\pi\rho|t|)^2 - x^2]/\pi((\pi\rho|t|)^2 + x^2)^2 \quad . \tag{5.6}$$

For t=0 the spatial decay is $-x^{-2}$. The small k limit washes out the oscillations. For x=0 the time decay is t^{-2}. This should be compared with a system with short range forces, where the decay in one dimension is $t^{-1/2}$. The t^{-2} reflects therefore dynamically the stiffness of the system.

The scaling of S(k,t) implies the scaling of the fluctuation field as

$$\xi^\varepsilon(f,t) = \sum_j f(\varepsilon x_j(\varepsilon^{-1}t)) - \rho\int dx f(\varepsilon x) \quad . \tag{5.7}$$

Note that there is <u>no</u> prefactor in the front of the sum. (5.5) translates into

$$\lim_{\varepsilon\to 0} \langle\xi^\varepsilon(f,t)\xi^\varepsilon(h,0)\rangle = \int dk \hat{f}(k)\hat{h}(-k)(|k|/2\pi)\exp[-|t||k|\pi\rho] \tag{5.8}$$

for test functions f,h such that $\int dk|k||\hat{f}(k)|^2<\infty$, $\int dk|k||\hat{h}(k)|^2<\infty$.

6. Scaling Limit : Fluctuation Field

What is the limit distribution of the fluctuation field $\xi^{\varepsilon}(f,t)$ as $\varepsilon \to 0$? By (5.7) $\xi^{\varepsilon}(f,t)$ counts the excess density in a large region of size ε^{-1}. This excess density is apparently of order 1. We are therefore far from the setting of a sum of weakly dependent random variables. To my own surprise the limit field turns out to be Gaussian provided the test functions are not too rough.

Theorem. Let $\xi^{\varepsilon}(f,t)$ be the fluctuation field defined by (5.7) with $t \in R$ and f such that

$$\int dk|k||\hat{f}(k)| < \infty \;, \quad \int dk|\hat{f}(k)| < \infty \;, \quad \int dk|\hat{f}(k)|^2 < \infty \;. \tag{6.1}$$

Then in law

$$\lim_{\varepsilon\to 0} \xi^{\varepsilon}(f,t) = \xi(f,t) \tag{6.2}$$

in the sense of convergence of joint moments. The limit field $\xi(f,t)$ is Gaussian with mean zero and covariance (5.8).

Remark. $\xi(f,t) = \int dx f(x)\xi_t(x)$ is the stationary Ornstein-Uhlenbeck process governed by

$$d\xi_t = -A\xi_t dt + B dW_t \;. \tag{6.3}$$

Here $A=\rho\pi(-d^2/dx^2)^{1/2}$ and $B=\sqrt{\rho}\,d/dx$ as linear operators and dW_t is space-time white noise.

We first prove (6.2) for $t=0$.

Proposition 5. In the sense of convergence of joint moments

$$\lim_{\varepsilon\to 0} \xi^{\varepsilon}(f) = \xi(f) \;. \tag{6.4}$$

$\xi(f)$ is Gaussian with mean zero and covariance

$$\langle\xi(f)\xi(h)\rangle = \int dk|k|\hat{f}(k)\hat{h}(-k) \;. \tag{6.5}$$

Lemma 1. Let χ_{ε} be the indicator function of the set $[-\varepsilon^{-1}\pi\rho, \varepsilon^{-1}\pi\rho]$, $\nabla = d/dx$. Let f be rapidly decreasing at infinity. Then

$$\chi_{\varepsilon}(-i\nabla)(e^{f} - 1) \tag{6.6}$$

considered as a linear operator on $L^2(R,dx)$ is of trace class. We have

$$<\exp(\sum_j f(\varepsilon x_j))> = \det(1 + \chi_\varepsilon(-i\nabla)(e^f - 1)) \quad . \tag{6.7}$$

Proof : Let $\{a^+_j, a_j | j = 1,\ldots,N\}$ be a finite-dimensional CAR-algebra. Let A,B,C be NxN matrices. If

$$e^A e^B = e^C \quad , \tag{6.8}$$

then

$$\exp[(a^+,Aa)]\exp[(a^+,Ba)] = \exp[(a^+,Ca)] \quad , \tag{6.9}$$

where (a^+,Aa) is shorthand for $\sum_{i,j} a^+_i A_{ij} a_j$.

We have

$$\exp[(a^+,Aa)]a^+(h)\exp[-(a^+,Aa)] = a^+(e^A h)$$

and

$$\exp[-(a^+,Aa)]|0> = |0> \tag{6.10}$$

where $|0>$ is the vacuum vector. Now (6.9) applied to an arbitrary vector yields, using (6.8),

$$\exp[(a^+,Aa)]\exp[(a^+,Aa)]\prod_j a^+(h_j)|0> = \prod_j a^+(e^A e^B h_j)|0> =$$

$$= \exp[(a^+,Ca)] \prod_j a^+(h_j)|0> \quad . \tag{6.11}$$

Since

$$tr[\exp[(a^+,Aa)]] = \det(1 + e^A) \quad , \tag{6.12}$$

$$tr[\exp[-\beta(a^+,Ha)]\exp[(a^+,Aa)]]/tr[\exp[-\beta(a^+,Ha)]]$$

$$= \det(1 + (e^{\beta H}+1)^{-1}(e^A-1)) \quad . \tag{6.13}$$

Let $H = H^*$ and let $\chi_-(H)$ be the spectral projection of H for the interval $(-\infty,0)$ and $\chi_o(H)$ the one for $\{0\}$. Let $< >_H$ denote the ground state expectation for (a^+,Ha). Taking the limit $\beta\to\infty$ in (6.13) yields

$$<\exp[(a^+,Aa)]>_H = \det(1 + (\chi_-(H)+\tfrac{1}{2}\chi_o(H))(e^A-1)) \quad . \tag{6.14}$$

We expand

$$\langle \exp\,(a^+,Aa)\rangle_H = \sum_{n=0}^{\infty} \frac{1}{n!} \langle (a^+,Aa)^n \rangle_H \quad . \tag{6.15}$$

The n-th moment is given by the Plemelj-Smithies formulae [15,16].

To prove (6.7) we expand and use Proposition 2,

$$\langle \exp[\int dx f(x) a^+(x) a(x)]\rangle = \sum_{n=0}^{\infty} \frac{1}{n!} \langle (\int dx f(x) a^+(x) a(x))^n \rangle \quad . \tag{6.16}$$

By the argument above, the n-th coefficient is given by the Plemelj-Smithies formulae

$$\det \begin{vmatrix} \mathrm{tr}\,B & n-1 & \dots & 0 \\ \mathrm{tr}\,B^2 & \mathrm{tr}B & \dots & 0 \\ \cdot & \cdot & & \\ \cdot & \cdot & & \\ \cdot & \cdot & & \\ \mathrm{tr}\,B^n & \mathrm{tr}B^{n-1} & \dots & \mathrm{tr}B \end{vmatrix} \tag{6.17}$$

with

$$B = \chi_\varepsilon(-i\nabla)(e^f - 1) \quad . \tag{6.18}$$

Since B is of trace class, the claim follows ∎

<u>Lemma 2.</u> Let

$$\phi_\varepsilon(z) = -\log\langle \exp[z \sum_j f(\varepsilon x_j)]\rangle + \varepsilon^{-1}\rho\int dx z f(x) \quad . \tag{6.19}$$

Then there exists a z_o independent of ε, $z_o > 0$, such that for $|z| < z_o$ $\phi_\varepsilon(z)$ has a convergent power series expansion and such that for $|z| < z_o$

$$\lim_{\varepsilon\to 0} \phi_\varepsilon(z) = \phi(z) = \sum_{n=1}^{\infty} z^{n+1} \frac{1}{(n+1)!} \int dk_1 \dots dk_n \hat{f}(k_1)\dots\hat{f}(k_n)\hat{f}(-k_1-\dots-k_n)$$
$$\frac{1}{2\pi}(S(k_1,\dots,k_n) + S(-k_1,\dots,-k_n)) \quad . \tag{6.20}$$

The function S is defined in (6.32) below.

Proof : If

$$\sup_x |e^{zf(x)} - 1| < 1 \quad , \tag{6.21}$$

then by Lemma 1 and by [16, VIII.17, Lemma 6]

$$\phi_\varepsilon(z) = \sum_{n=1}^{\infty} (-1)^n \frac{1}{n} \operatorname{tr}[(\chi_\varepsilon(-i\nabla)(e^{zf} - 1))^n] + \varepsilon^{-1}\rho\int dxzf(x) \tag{6.22}$$

is convergent. Let $e^{zf} - 1 = h$. Then

$$\operatorname{tr}[(\chi_\varepsilon(-i\nabla)h)^{n+1}] = \frac{1}{2\pi}\int d\lambda\int dk_1 \ldots dk_n \chi_\varepsilon(\lambda)\chi_\varepsilon(\lambda+k_1)\ldots$$

$$\chi_\varepsilon(\lambda+k_1+\ldots+k_n)\hat{h}(k_1)\ldots\hat{h}(k_n)\hat{h}(-k_1-\ldots-k_n) \quad . \tag{6.23}$$

We integrate over λ. For fixed k and ε sufficiently small this yields

$$\rho\varepsilon^{-1} - \frac{1}{2\pi}(\max(0,k_1,\ldots,k_1+\ldots+k_n) + \max(0,-k_1,\ldots,-k_1\ldots-k_n)). \tag{6.24}$$

We resum the term proportional to ε^{-1}. This gives

$$\varepsilon^{-1}\rho \sum_{n=1}^{\infty} (-1)^n \frac{1}{n}\int dxh(x)^n = -\varepsilon^{-1}\rho\int dxzf(x) \tag{6.25}$$

and cancels the second term in (6.22).

We order in powers of z by expanding

$$e^{zf} - 1 = \sum_{n=1}^{\infty} \frac{1}{n!}(zf)^n \quad . \tag{6.26}$$

Then a typical term of order z^{n+1} is

$$\operatorname{tr}[\chi_\varepsilon(-i\nabla)f^{n_1}\ldots\chi_\varepsilon(-i\nabla)f^{n_m}] \tag{6.27}$$

with the constraints $n_j \geqq 1$ and

$$\sum_{j=1}^{m} n_j = n+1 \quad . \tag{6.28}$$

We integrate as in (6.23) which gives

$$\frac{1}{2\pi}\int d\lambda\int dk_1\ldots dk_n\chi_\varepsilon(\lambda)\chi_\varepsilon(\lambda+k_1+\ldots+k_{n_1})\ldots\chi_\varepsilon(\lambda+k_1+\ldots+k_{n_1+\ldots+n_{m+1}})$$

$$\hat{f}(k_1)\ldots\hat{f}(k_n)\hat{f}(-k_1-\ldots-k_n) \quad . \tag{6.29}$$

We symmetrize through all possible permutations of the indices of the k's. The number of such permutations is, using (6.28),

$$n!n_m \prod_{j=1}^{m} \frac{1}{n_j!} \quad . \tag{6.30}$$

We expand now in (6.22). Let us consider the terms with given $(n_1,\ldots,n_m)$ together with their cyclic permutations $(n_m,n_1,\ldots,n_{m-1})$, etc. . Let p be the number of permutations which are identical. Then the contribution is

$$z^{n+1}(-1)^m \frac{1}{mp} \prod_{j=1}^{m} \frac{1}{n_j!}\{\mathrm{tr}[\prod_{j=1}^{m} \chi_\varepsilon(-i\nabla)f^{n_j}] + (m-1) \text{ cyclic permutations}\}$$

$$= z^{n+1}(-1)^m \frac{1}{p} \prod_{j=1}^{m} \frac{1}{\alpha_j!} \mathrm{tr}[\prod_{j=1}^{m} \chi_\varepsilon(-i\nabla)f^{n_j}]$$

$$= z^{n+1}(-1)^m \frac{1}{(n+1)!} n! \prod_{j=1}^{m} \frac{1}{n_j!}\{n_m \mathrm{tr}[\prod_{j=1}^{m} \chi_\varepsilon(-i\nabla)f^{n_j}]$$

$$+ ((m/p) - 1) \text{ cyclic permutations}\} \quad . \tag{6.31}$$

Here we sum only over permutations which differ from each other.

To define the limit $\varepsilon \to 0$ and for later purposes it is convenient to introduce some notation. Let us denote by $\alpha = (\alpha_1,\ldots,\alpha_n),\beta,\ldots,$ vectors with entries $\alpha_j = 0,1$. $\alpha < \beta$ is understood coordinate-wise, i. e. $\alpha_j \leqq \beta_j$ with strict inequality for at least one j. Such vectors $\alpha, \alpha \neq 0$, are called <u>branches</u>. A set of ordered branches $\alpha^{(1)} <\ldots< \alpha^{(m)}$ is called a <u>tree</u>, generically denoted by T. $|T|$ is the number of branches of T, $|T| \leqq n$. $\alpha \in T$ means that α is a branch of the tree T. $\mathcal{T}(n)$ is the set of all trees consisting of branches formed with n-vectors. We define the function

$$S(k_1,\ldots,k_n) = - \sum_{T, T\in \mathcal{T}(n)} (-1)^{|T|} \max\{0, \alpha\cdot k \mid \alpha \in T\} \quad . \tag{6.32}$$

$\max\{0,\alpha\cdot k \mid \alpha \in T\}$ is called the maximum of the tree T.

We return to (6.31) whose terms are precisely exhausted by choosing all possible permutations in (6.29). These are labeled by trees. The m-1 branches of the basic tree are given by $\alpha_j^{(\ell)} = 1$ for $j = 1,\ldots, n_1+\ldots+n_\ell$ and $\alpha_j^{(\ell)} = 0$ otherwise, $\ell = 1,\ldots,m-1$. The other trees are obtained from permutations of this basic tree.

In the limit $\varepsilon \to 0$, the terms of order ε^{-1} cancel each other, as shown already, and the remainder yields (6.20).

We still have to determine the new radius of convergence. In

$(6.29)_n$ we substract the dominant term $2\pi\rho\varepsilon^{-1}$. The remainder is bounded by $\sum_{j=1}^{n} |k_j|$ independently of ε. This yields the bound

$$n\int dk|\hat{f}(k)||k|(\int dk|\hat{f}(k)|)^{n-2}\int dk|\hat{f}(k)|^2 \quad . \tag{6.33}$$

Inserting this back into the expansion shows that the radius of convergence is determined by

$$\exp[|z|\int dk|\hat{f}(k)|] < 2 \quad \blacksquare \tag{6.34}$$

Lemma 3. We have

$$S(k_1) + S(-k_1) = |k_1| \tag{6.35}$$

and

$$S(k_1,\ldots,k_n) = 0 \quad , \quad n \geqq 2 \quad .$$

Proof : The proof uses an induction argument. One checks that

$$S(k_1,k_2) = 0 \quad . \tag{6.37}$$

Let me assume then that

$$S(k_1,\ldots,k_n) = 0 \tag{6.38}$$

for some $n \geqq 2$. We want to show that for all $y \in R$

$$S(y,k_1,\ldots,k_n) = 0 \quad . \tag{6.39}$$

Because S is symmetric we may order $y = k_o < k_1 \ldots < k_n$. We write $k = (k_1,\ldots,k_n)$.

Let us choose a k such that

$$\alpha\cdot k \neq \alpha'\cdot k \ , \ \alpha\cdot k \neq 0 \ , \tag{6.40}$$

for arbitrary $\alpha \neq \alpha'$, cf. (6.32) above for the definition of α,α'. We call such a k nondegenerate. Since S is continuous, S at degenerate points follows. Let us consider the function

$$y \longmapsto S(y,k) \quad . \tag{6.41}$$

$S(\cdot,k)$ is piecewise linear and $S(y,k)$ is zero for y sufficiently negative, by the induction assumption. $S(\cdot,k)$ can change its slope only at points of degeneracy of (y,k), i.e. whenever

$$\alpha_o y + \alpha\cdot k = \alpha_o' y + \alpha'\cdot k \quad , \tag{6.42}$$

$\alpha_o, \alpha_o' = 0,1$.

If either $\alpha_o = 1 = \alpha_o'$ or $\alpha_o = 0 = \alpha_o'$, then (6.42) has no solution by the assumption on k. (6.42) admits then the solutions

$$y + \alpha\cdot k = \alpha'\cdot k \quad . \tag{6.43}$$

If $T \in \mathcal{T}(n+1)$ does not contain a branch $(1,\alpha)$ for any α, then the maximum of T does not depend on y. Suppose then $(1,\alpha) \in T$ for some $\alpha, \alpha = 0$ included. If $(1,\alpha')$, $\alpha' \neq 0$, is not a branch of T, then the slope of maximum of T does not change under a small change of y. For T to contain both branches $(1,\alpha)$ and $(0,\alpha')$ we must have $\alpha' \leqq \alpha$.

We conclude : $y \longmapsto S(y,k)$ can change its slope only at points

$$y + \alpha\cdot k = 0 \quad , \tag{6.44}$$

including $\alpha=0$. The degeneracy points of (y,k) are then precisely

$$y + \beta\cdot k = (\beta - \alpha)\cdot k \tag{6.45}$$

with arbitrary $\beta \geqq \alpha$.

We fix now some α and y defined through (6.44). We assume that for $y' < y$ $S(y',k) = 0$. We want to determine the change

$$\Delta S = S(y+\varepsilon,k) - S(y,k) \quad , \tag{6.46}$$

ε small. Let us consider one tree, T, out of the sum for S and call its change Δ_T. We have two possibilities. 1.) T does not contain both branches $(1,\beta)$ and $(0,\beta-\alpha)$ for any $\beta > \alpha$. Then $\Delta_T=0$. For some β with $\beta>\alpha$ $(1,\beta),(0,\beta-\alpha) \in T$ and $y + \beta\cdot k$ is not the maximum of T. Then a small change in y does not alter the situation and therefore $\Delta_T = 0$. In both cases we define

$$\Delta(T) = 0 \quad . \tag{6.47}$$

2.) For some $\beta, \beta \geq \alpha$, $(1,\beta), (0,\beta-\alpha) \in T$ and $y+\beta\cdot k$ is the maximum of T (which includes zero if $\beta-\alpha=0$). Then

$$\Delta_T = y + \varepsilon + \beta\cdot k - (\beta-\alpha)\cdot k = \varepsilon \quad . \tag{6.48}$$

In this case we define

$$\Delta(T) = 1 \quad . \tag{6.49}$$

Altogether we conclude

$$\Delta S = \varepsilon \sum_{T, T \in \mathcal{T}(n+1)} (-1)^{|T|} \Delta(T) \quad . \tag{6.50}$$

The goal is to show that $\Delta S = 0$ which proves then the lemma. We have to build a cancellation scheme which exploits

$$\sum_{T, T \in \mathcal{T}(m)} (-1)^{|T|} = 0 \tag{6.51}$$

for m=1,2,..., where the summation includes the empty tree.

A tree T cannot have two pair of branches $(1,\beta)$, $(0,\beta-\alpha)$ and $(1,\beta')$, $(0,\beta'-\alpha)$ with $\beta \neq \beta'$. Therefore the sum (6.50) may be restricted to all T's such that $(1,\beta)$, $(0,\beta-\alpha) \in T$ with fixed $\beta \geq \alpha$

Summation scheme.

We consider trees $T \in \mathcal{T}(m)$. Let $\alpha, \alpha \neq 0$, be given, not necessarily the α from above. We also assume the branches larger than α to be fixed without explicit mentioning. The set of all such trees is denoted by $\mathcal{T}(\alpha)$. We decompose

$$\alpha = \gamma + \delta \quad , \tag{6.52}$$

where $\gamma_j = 0$ whenever $k_j > 0$ and $\delta_j = 0$ whenever $k_j < 0$, j = 1,... m. We use γ and δ as generic notation for this decomposition. (α) is divided recursively into classes

$$\mathcal{T}(\gamma^{(1)}; \delta^{(1)}, \ldots, \delta^{(\ell)}) . \tag{6.53}$$

These are mutually exclusive and together exhaust $\mathcal{T}(\alpha)$.

(i) If either $\alpha = \gamma$ or $\alpha = \delta$, there is only one class denoted again by $\mathcal{T}(\alpha)$.

We assume then $\gamma \neq 0$ and $\delta \neq 0$. Consider the smallest branch, β, of $T \in \mathcal{T}(\alpha)$, which could be α itself.

(ii) $\beta = \gamma^{(1)} + \delta^{(1)}$, $\gamma^{(1)} \neq 0$, $\delta^{(1)} \neq 0$. Then T belongs to $\mathcal{T}(\gamma^{(1)};\delta^{(1)})$.

(iii) $\beta = \bar{\gamma}$ and $\gamma^{(1)} + \delta^{(1)}$, $\delta^{(1)} \neq 0$, is the smallest branch larger than β. Then T belongs to $\mathcal{T}(\gamma^{(1)};\delta^{(1)})$.

(iv) $\beta = \delta$. Then T belongs to $\mathcal{T}(0;\delta)$.

(v) $\beta = \delta^{(1)}$ and $\delta^{(1)} \neq \delta$. We then delete $\delta^{(1)}$ from every branch and go through steps (ii) to (v). This determines the classes $\mathcal{T}(\gamma^{(1)};\delta^{(1)},\delta^{(2)})$ with $\gamma^{(1)} = 0$ included, etc..

From (6.51) we deduce

$$\sum_{T \in \mathcal{T}(\gamma^{(1)};\delta^{(1)},\ldots,\delta^{(\ell)}),\gamma^{(1)} \neq 0} (-1)^{|T|} = 0 . \qquad (6.54)$$

Cancellations

1.) $\beta = \alpha$

1.1) $\alpha = 0$. By (6.51) the sum over all branches larger than (1,0) cancel.

1.2) $\alpha \neq 0$. We decompose $(1,\alpha) = \gamma + \delta$. Because $y + \alpha \cdot k = 0$ necessarily $\gamma \neq 0, \delta \neq 0$. Let $T \in \mathcal{T}(\gamma^{(1)};\delta^{(1)})$. If for some branch $\beta' \in T$, $\beta' \neq (1,\alpha)$, $\beta' \geq \gamma^{(1)} + \delta^{(1)}$, we have $\beta' \cdot (y,k) > 0$, then $\Delta(T) = 0$. If for all branches $\beta' \in T$, $\beta' \neq (1,\alpha)$, $\beta' \geq \gamma^{(1)} + \delta^{(1)}$, we have $\beta' \cdot (y,k) < 0$, then the maximum of the tree is at $y + \alpha \cdot k$, independently of the choice of T, because $\beta'' \cdot (y,k) < 0$ for all $\beta'' \leq \gamma^{(1)} + \delta^{(1)}$. By (6.54) the partial sum is zero.

Let $T \in \mathcal{T}(\gamma;\delta^{(1)},\ldots,\delta^{(\ell)})$, $\ell \geq 2$, or $T \in \mathcal{T}(0;\delta^{(1)},\ldots,\delta^{(\ell)})$. Then the smallest branch of T is positive which implies $\Delta(T) = 0$.

We conclude $\Delta S = 0$.

2.) $\beta > \alpha$. Since $(0,\beta-\alpha) < (1,\beta)$, we consider $(0,\beta-\alpha)$. We decompose $(0,\beta-\alpha) = \gamma + \delta$.

2.1) $\gamma \neq 0, \delta \neq 0$. Let $T \in \mathcal{T}(\gamma^{(1)};\delta^{(1)},\ldots,\delta^{(\ell)})$. If $(\beta-\alpha)\cdot k$ is not the maximum of T, then $\Delta(T) = 0$. If $(\beta-\alpha)\cdot k$ is the maximum of T, then $(\gamma^{(1)} + \Sigma\delta^{(j)})\cdot(y,k) < (\beta-\alpha)\cdot k$ independently of the choice of T. By (6.54) the partial sum is zero. The only class left from $\mathcal{T}(\beta-\alpha)$ is $\mathcal{T}(0;\delta^{(1)},\ldots,\delta^{(\ell)})$ with $\delta^{(1)} + \ldots + \delta^{(\ell)} = \delta$. By construction $\delta\cdot k > (\beta-\alpha)\cdot k$ and therefore for every tree in this class $\Delta(T) = 0$.

2.2) $\gamma \neq 0, \delta = 0$. Then $(\beta-\alpha)\cdot k < 0$ and the maximum of the tree has to be either at some other branch or zero. Therefore $\Delta(T) = 0$.

2.3) $\gamma = 0, \delta \neq 0$. Every branch smaller than $(1,\beta)$ and larger than

$\beta-\alpha = \delta$ contains δ. We delete from every branch the segment δ and fix the branches smaller than δ. This leaves branches smaller than $(1,\alpha)$. We partially resum to zero as in 1.). The trees left must have a branch $\delta'+\delta$, $\delta' \neq 0$. Clearly $\delta'\cdot k + \delta\cdot k > \delta\cdot k$ and therefore $\Delta(T) = 0$ ∎

Proof of Proposition 5 : Under the condition (6.1) the generating function has a finite radius of convergence independent of ε, cf. (6.33), and has a limit as $\varepsilon\to 0$ by Lemma 2. This implies that the moments converge. By Lemma 3 the limiting moments are the moments of the Gaussian measure with covariance (6.5) ∎

Proof of the Theorem : We reduce the convergence of the time-dependent moments to the one of the static moments. The construction is carried through for two times, but the argument is general. The moments are organized somewhat differently than before.

Step 1. Let $\{p(1),\ldots,p(n)\}$ be a permutation of $\{1,\ldots,n\}$. sign(p) is the sign of the permutation. P is the set for all permutations. The dependence on n will be clear from the context. $P^* \subset P$ is the set of all permutations such that $p(j) \neq j$, $j = 1,\ldots,n$, P_g is the set of all Gaussian permutations. If n is odd, P_g is empty. If n is even, P_g are all those permutations which can be ordered in pairs, i.e. pairs $\{i,j\}$ with $i = p(j)$, $j = p(i)$.

We have

$$< \prod_{j=1}^{n} \xi^{\varepsilon}(f_j) > = \int \prod_{j=1}^{n} dk_j \hat{f}(k_j) S^{\varepsilon}(k_1,\ldots,k_n) \tag{6.55}$$

with

$$S^{\varepsilon}(k_1,\ldots,k_n) = \sum_{p\in P^*} \text{sign}(p) \int \{ \prod_{j=1}^{n} dx_j \delta(k_j+x_j-x_{p(j)}) \}$$

$$\prod_{j<p(j)} \chi_{\varepsilon}(x_{p(j)}) \prod_{j>p(j)} (\chi_{\varepsilon}-1)(x_{p(j)}) \tag{6.56}$$

Equalities for S^{ε}, S^{ε}_{ir}, etc., will be understood in the distributional sense, i.e. when integrating over smooth test functions with compact support.

For given p we connect the points $\{1,\ldots,n\}$ by directed bonds $\{j \text{ to } p(j) \mid j=1,\ldots,n\}$. We call a permutation irreducible if the loop closes only after n steps. The set of all irreducible permutations is denoted by P_{ir}. If n is even, the Gaussian permutations are those which factor into n/2 loops of lenth two. We define

$$S^{\varepsilon}_{ir}(k_1,\dots,k_n) = (-1)^{n+1} \sum_{p\in P_{ir}} \int\{\prod_{j=1}^{n} dx_j \delta(k_j+x_j-x_{p(j)})\}$$

$$\prod_{j<p(j)} \chi_\varepsilon(x_{p(j)}) \prod_{j>p(j)} (\chi_\varepsilon-1)(x_{p(j)}) \quad . \tag{6.57}$$

(6.56) can be factored into irreducible parts. For n=2 the limit $\varepsilon\to 0$ gives the covariance (6.5). From Proposition 5 we conclude that

$$\lim_{\varepsilon\to 0} S^{\varepsilon}_{ir}(k_1,\dots,k_n) = 0 \tag{6.58}$$

for $n\geq 3$.

Step 2. We repeat the construction for the time-dependent moments. This yields

$$< \prod_{j=1}^{m} \xi^{\varepsilon}(f_j,t) \prod_{j=m+1}^{n} \xi^{\varepsilon}(f_j)> = \int \prod_{j=1}^{n} dk_j \hat{f}(k_j) S^{\varepsilon}(k_1,\dots,k_m,t;k_{m+1},\dots,k_n) \tag{6.59}$$

with

$$S_{ir}(k_1,\dots,k_m,t;k_{m+1},\dots,k_n) = (-1)^{n+1} \sum_{p\in P_{ir}} \int\{\prod_{j=1}^{n} dx_j$$

$$\delta(k_j+x_j-x_{p(j)})\} \prod_{j=1}^{m} \exp[\ \varepsilon t((k_j+x_j)^2-x_j^{\,2})/2]$$

$$\prod_{j<p(j)} \chi_\varepsilon(x_{p(j)}) \prod_{j>p(j)} (\chi_\varepsilon-1)(x_{p(j)}) \quad . \tag{6.60}$$

The first m arguments refer to $t>0$ and the remaining ones to $t=0$, $1\leq m\leq n-1$. For n=2 and m=1 the limit $\varepsilon\to 0$ in (6.60) yields the covariance (5.8). To prove the Theorem we have to show then that

$$\lim_{\varepsilon\to 0} S^{\varepsilon}_{ir}(k_1,\dots,k_m,t;k_{m+1},\dots,k_n) = 0 \tag{6.61}$$

for $n\geq 3$, $1\leq m\leq n-1$.

Step 3. We introduce a labeling of the integral associated to an irreducible permutation. To each directed bond $j\to p(j)$ we assign the number k_j. To the site at the endpoint of the directed bond we assign either a χ_ε, if the bond is directed forward, or a $\chi_\varepsilon-1$, if the bond is directed backward. We now set $x_{p(n)}=-\lambda$. The site p(n) is regarded as the starting point of the closed loop. We successively integrate out the x's following the direction of the loop. This yields an integral of the form

$$[\int d\lambda \prod_{\ell}(\chi_\varepsilon -1)(\lambda+\beta^{(\ell)}\cdot k)\prod_{\ell}\chi_\varepsilon(\lambda+\alpha^{(\ell)}\cdot k)]\,\delta(k_1+\ldots+k_n) \quad . \tag{6.62}$$

Here $k = (k_1,\ldots,k_{n-1})$ and $\alpha^{(\ell)},\beta^{(\ell)}$ are $(n-1)$-vectors with entries 0,1. They depend only on the permutation p and are defined as follows.

By construction the starting point of the loop has assigned to it a $\chi_\varepsilon-1$. This gives in the integrand the factor $(\chi_\varepsilon-1)(\lambda)$ and $\beta^{(1)}=0$. Suppose we arrive at the m-th point of the loop. Then the integrand picks up either a χ_ε or a $\chi_\varepsilon-1$ whichever is assigned to the m-th point of the loop. The argument of this function is either $\lambda+\alpha^{(\ell)}\cdot k$ or $\lambda+\beta^{(\ell)}\cdot k$, where the linear combination of the k's is defined by summing along the loop from the starting point to the m-th point. By construction the integrand depends only on $k_1,\ldots,k_{n-1}$. Because there is at least one forward and one backward loop the integrand contains at least one χ_ε and one $\chi_\varepsilon-1$ factor.

In (6.60) we pick a specific p and fix k. This gives then an integral of the form, $a < b,c$ are coefficients depending linearly on k,

$$\lim_{\varepsilon\to 0}\int_{\pm\varepsilon^{-1}\pi\rho+a}^{\pm\varepsilon^{-1}\pi\rho+b} d\lambda\ \exp[-\varepsilon tc\lambda]\exp[O(\varepsilon)] = (b-a)\ \exp[(\pm)t\pi\rho c] \quad . \tag{6.63}$$

Let me define the limit more explicitely. For specified irreducible permutation p the vectors $\alpha^{(\ell)},\beta^{(\ell)}$ are defined above. Let then

$$\begin{aligned} a_+ &= \max_{\ell}\{\alpha^{(\ell)}\cdot k\}, \quad a_- = \min_{\ell}\{\alpha^{(\ell)}\cdot k\} \\ b_+ &= \max_{\ell}\{\beta^{(\ell)}\cdot k\}, \quad b_- = \min_{\ell}\{\beta^{(\ell)}\cdot k\} \end{aligned} \tag{6.64}$$

Note that we cannot satisfy $a_- > b_+$ and $b_- > a_+$ simultaneously.

We use (6.63) to obtain the contribution of p to (6.60) in the limit $\varepsilon \to 0$.

(i) If $a_- < b_+$ and $b_- < a_+$ the contribution is zero.

(ii) If $a_- > b_+$, then the contribution is

$$(a_- - b_+)\exp[-t\pi\rho \sum_{j=1}^{m} k_j] \quad . \tag{6.65}$$

(iii) If $b_- > a_+$, then the contribution is

$$(b_- - a_+)\exp[t\pi\rho \sum_{j=1} k_j] \quad . \tag{6.66}$$

Since (ii) and (iii) exclude each other, we write the limit $\varepsilon \to 0$ of (6.60) as

$$S_{ir}(k_1,\ldots,k_m,t;k_{m+1},\ldots,k_n) = \delta(k_1+\ldots+k_n)$$
$$\{\exp[-t\pi\rho \sum_{j=1}^{m} k_j]S_+(k) + \exp[t\pi\rho \sum_{j=1}^{m} k_j]\, S_-(k)\} \quad . \tag{6.67}$$

Schwarz's inequality in (6.59) shows that S_{ir} has to be bounded in t. Let $m = n-1$. We conclude, by choosing t large enough, that $S_+(k) = 0$ whenever $\sum_{j=1}^{n-1} k_j < 0$ and $S_-(k) = 0$ whenever $\sum_{j=1}^{n-1} k_j > 0$. Therefore S_+ and S_- have disjoint support. By (6.58)

$$S_+(k) + S_-(k) = 0 \tag{6.68}$$

for $n \geqq 3$. This is possible only if S_+ and S_- vanish separately. Therefore $S_+ = 0 = S_-$ which proves (6.61) by (6.67) ∎

7. An Interface Model from Statistical Mechanics

The connection between the infinitely strong repulsive δ-Bose gas and the ideal Fermi gas in one dimension is ancient, cf.[14] for the history and [17] for a more recent development.

De Gennes [18] uses this connection in an application in spirit close to our study. He considers N polymers in solution constrained to a plane and not allowed to turn "backward". This is modeled as N one-dimensional random walks which are mutually self-avoiding at any given time. Each allowed configuration has equal weight. Let us translate to Brownian motion : We consider N independent Brownian motions on the circle [-L,L]. They start at time -T, say regularly spaced, and run up to time T. We condition the Brownian motions not to touch each other and take the limit $T\to\infty$. According to Section 4 we thereby obtain a stationary stochastic process corresponding to N Brownian particles interacting through the pair force $(\pi/2L)\cot(\pi x/2L)$. The unusual feature of this model is that a quantum mechanical pair potential (i.e. the conditioning of each pair of Brownian motions not to touch) translates into a pair force. In general, following such a procedure, one ends up with a many-body force which has no natural interpretation in terms of interacting Brownian particles.

De Gennes' model has been revived (and extended to higher dimensions) in the context of commensurate-incommensurate phase transitions

[19,20,21].

As an example let us consider the two-dimensional ANNNI model which has a zero temperature phase consisting of stripes $\begin{smallmatrix} ++--++--\cdots \\ ++--++--\cdots \\ ++--++--\cdots \end{smallmatrix}$. At low temperatures the interfaces separating +- and -+ fluctuate with a statistical weight inherited from the Gibbs measure. In approximation one may think of the interfaces as random walks (time running upward) with no touching allowed.

Our contribution here is that

(i) the model of random walks conditioned not to touch has a natural interpretation as a system of Brownian particles interacting with the pair force $1/x$,

(ii) viewed from this angle the model has an unusual hydrodynamic behavior for which so far no rigorous example had been known.

References

[1] M.L. Mehta, Random Matrices and the Statistical Theory of Energy Levels, Academic Press, New York 1967

[2] F.J. Dyson, J.Math.Phys. 3, 1191 (1962)

[3] R. Lang, Z. Wahrscheinlichkeitstheorie Verw.Geb. 38, 55 (1977)

[4] H. Spohn, Comm.Math.Phys. 103, 1 (1986)

[5] K. Gawedzki, A Kupiainen, Comm.Math.Phys. 92, 531 (1984)

[6] J. Magnen, R. Séneor, preprint, Ecole Polytechnique

[7] H. Spohn, to be published

[8] J.-R. Fontaine, Comm.Math.Phys. 91, 419 (1983)

[9] R. Holley and D.W. Stroock, Ann.Math. 110, 333 (1979)

[10] E. Presutti and H. Spohn, Ann.Prob. 11, 867 (1983)

[11] O.E. Lanford, Comm.Math.Phys. 11, 257 (1969)

[12] A. Lenard, Comm.Math.Phys. 30, 35 (1973)

[13] O. Bratelli and D.W. Robinson, Operator algebras and quantum statistical mechanics, Vol.I and II. Springer, Berlin, 1981

[14] E. Lieb and D. Mattis, Mathematical Physics in One Dimension, Academic Press, New York, 1966

[15] B. Simon, Trace ideals and their applications, London Mathematical Society Lecture Note Series Vol. 35, Cambridge University Press 1979

[16] M. Reed, B. Simon, Methods of modern mathematical physics, Vol.IV Academic Press, New York, 1978

[17] E. Witten, Comm.Math.Phys. 92, 455 (1984)

[18] P.G. De Gennes, J.Chem.Phys. 48, 2257 (1968)

[19] J. Villain and P. Bak, J.Physique 42, 657 (1981)

[20] T.W. Burkhardt and P. Schlottmann, Z.Physik B54, 151 (1984)

[21] J. Bricmont, A. El Mellouki and J. Fröhlich, J.Stat.Phys. 42, 743 (1986)

A PROPAGATION OF CHAOS RESULT FOR BURGERS' EQUATION

A.S. Sznitman

Universite Paris VI
Laboratoire de Probabilités
4 Place Jussieu Tour 56
3ème Etage
Associé C.N.R.S. n° 224
75252 Paris, Cedex 05, FRANCE

INTRODUCTION

In 1967, McKean [1] conjectured that if one considered a system of N particles on R with formal generator

$$L_N = \frac{1}{2} \sum_1^N \frac{\partial^2}{\partial x_i^2} + \frac{1}{2(N-1)} \sum_{i<j} \delta(x_i - x_j)\left(\frac{\partial}{\partial x_i} + \frac{\partial}{\partial x_j} \right), \tag{1.1}$$

this system would present a propagation of chaos property with respect to Burgers' equation:

$$\frac{\partial u}{\partial t} = \frac{1}{2} \frac{\partial^2 u}{\partial x^2} - \frac{1}{2} \frac{\partial}{\partial x} (u^2), \tag{1.2}$$

that is to say, starting with N particles at time 0, independent and u_0 distributed, if one fixes k, $t > 0$, when N goes to infinity, asymptotically the first k particles $x_t^1,\dots,x_t^k$ become independent and u_t distributed, if u_t is the solution of (1.2) with initial condition u_0. More recently there has been several results in this direction, by CALDERONI-PULVIRENTI [1], OELSCHLÄGER [1], introducing a smoothing $\phi_N(\cdot)$ of $\delta(\cdot)$, converging not too fast, and results by GUTKIN-KAC [1], KOTANI-OSADA [1], using analytical methods.

The intuition behind the conjecture relied on the fact that for a smooth function $b(\cdot)$, if one considers for each $N > 2$, the solutions $X_\cdot^1,\dots,X_\cdot^N$ of:

$$dX_t^i = dB_t^i + \frac{1}{2(N-1)} \sum_{j \neq i} b(X_t^i - X_t^j)dt, \quad i = 1,\dots,N, \tag{1.3}$$

$$X_0^i, \ u_0^{\otimes N} \text{ - distributed,}$$

$((B^i_\cdot)$, are independent real valued Brownian motions, independent of the initial conditions $(X^i_0))$.

The generator of the $(X^i_\cdot)$ is given by (1.1) with $\delta(\cdot)$ replaced by $b(\cdot)$, and if one constructs the (non linear) processes:

$$(1.4) \qquad d\overline{X}^i_t = dB^i_t + \frac{1}{2} \int b(\overline{X}^i_t - y)u_t(dy),$$

$\overline{X}^i_0 = X^i_0$, and $u_t(dy)$ the law of $\overline{X}^i_t$, (see McKEAN [1], TANAKA [1]), $X^{i,N}_\cdot \underset{N\to\infty}{\to} \overline{X}^i_\cdot$, and the time marginals of $\overline{X}^i$ evolve according to:

$$(1.5) \qquad \frac{\partial u}{\partial t} = \frac{1}{2} \frac{\partial^2 u}{\partial x^2} - \frac{1}{2} \frac{\partial}{\partial x} \left(\int b(x - y)u_t(dy)u\right).$$

Now if one lets formally $b(x - y) = \delta(x - y)$ in (1.5), one recovers Burgers' equation (1.2).

II. CONSTRUCTION OF THE N-PARTICLE SYSTEM

(for the details see SZNITMAN-VARADHAN [1]).

One considers the problem in R^d:

$$(2.1) \qquad dX_t = dB_t + \sum_{j=1}^{k} V_j \, dL^\circ(n_j \cdot X)_t,$$

where n_j are unit vectors defining distinct hyperplanes H_j

(2.20 $\qquad V_j$ are vectors in R^d satisfying $n_j \cdot V_j = 0$.

$(L^\circ(n_j \cdot X)$ denotes the symmetric local time in zero of $n_j \cdot X)$. If one uses formally $2 \cdot \delta(X^i_t - X^j_t)dt = dL^\circ(X^i - X^j)_t$, then choosing $n_{ij} = \frac{(e_i - e_j)}{\sqrt{2}}$, $V_{ij} = (e_i + e_j) \frac{\sqrt{2}}{4(n - 1)}$, $i < j$, one recovers formally the generator (1.1), as a special case of (2.1).

Let us now give some indications how one gets existence and uniqueness of solutions of (2.1) (trajectorially and in law) for an initial condition $X(0) \notin \bigcup_{j\neq i} H_j \cap H_i = S$.

If $T_1 < T_2 < \ldots < T_k \ldots$, denote the successive times of visit to different hyperplanes $H_{n_1}, H_{n_2}, \ldots, H_{n_k}$, of X, because of (2.2), for $t \leq T_{i+1} - T_i$ the solution of (2.1) looks like:

$$(2.3) \qquad X_{t+T_i} = X_{T_i} + (B_{t+T_i} - B_{T_i}) + V_{n_i} L^\circ(n_i \cdot (B_{\cdot+T_i} - B_{T_i}))_t$$

and one sees that the question of uniqueness (trajectorial and in law) of the solutions of (2.1) will be settled if the (T_i) do not accumulate. To this end, one uses the symmetries of the process.

When $d = 2$, roughly speaking, the idea is that if p denotes the inversion of pole zero and radius 1, and Y_n^x is the discrete time Markov chain starting from $x \in H \setminus S$, induced by the visits of the process X_t^x to successive hyperplanes, one has

$$(2.4) \qquad -p(Y_n^x) \underset{\text{law}}{\simeq} Y_n^{-p(x)},$$

moreover the probability of converging to zero can be seen to be independent of the starting point and is zero or one. Now by (2.4), it is equal to the probability of converging to infinity; both events being disjoint, this probability is zero.

From this fact one can deduce that the (T_i) do not accumulate.

When $d \geq 3$, one uses an other symmetry of the process. Namely, the process X construced with (n_i, V_i) is in duality with respect to Lebesgue measure, with the process $\hat{X}$ constructed with $(n_i, -V_i)$. This comes from the fact that the "vector fields" $\sum_i \delta(n_i \cdot x) V_i$ are divergence free, thanks to (2.2), so that formally the generator involved with (2.1) is $L = \frac{1}{2} \Delta + \sum_i \delta(n_i \cdot x) V_i \cdot \nabla$, and $L^* = \frac{1}{2} \Delta - \sum \delta(n_i \cdot x) V_i \cdot \nabla$, (KOTANI-OSADA [1], make this key remark on (1.1), and used P.D.E. results such as in ARONSON [1] to construct the process associated with (1.1)). By induction on the dimension one reduces the problem of accumulation of the (T_i), to a problem, of convergence of $X_\cdot$ to zero when t goes to T_∞. One can show that a.s. $A_{T_\infty} = \int_0^{T_\infty} \frac{ds}{|X_s|^2} = \infty$, and if (τ_t) denotes the inverse of (A_t), $Y_t = X_{\tau_t}$ and $\hat{Y}_t = \hat{X}_{\hat{\tau}_t}$ (with obvious notations) are truly Markovian and in duality with respect to $\frac{dx}{r^2}$, which since $d \geq 3$, is a Radon measure, and one sees easily that this prevents the convergence to zero. One can also show that for $x \notin S$:

(2.5) $$\lambda > 0, \quad \lambda X^{x}_{\cdot/\lambda^2} \simeq X^{\lambda x}_{\cdot}, \quad \text{and}$$

(2.6) $$\text{if} \quad dX^{\alpha}_t = dB_t + \sum_i \phi_\alpha(n_i \cdot X^{\alpha}_t) V_i dt, \; X^{\alpha}_0 = x$$

$$\text{for} \quad \phi_\alpha(\cdot) = \frac{1}{\alpha}\phi\left(\frac{\cdot}{\alpha}\right) \underset{\alpha\to 0}{\to} \delta(\cdot), \text{ then } X^{\alpha,x}_{\cdot} \underset{\alpha\to 0}{\to} X^x, \text{ a.s.}$$

III. THE PROPAGATION OF CHAOS RESULT

(for the details, see SZNITMAN [2]).

Let us start with the

Definition: Let E be a separable metric space, $v \in M(E)$, a sequence (v_N) of symmetric probabilities on E^N is said v-chaotic if:

(3.1) $$\text{for} \quad \phi_1,\ldots,\phi_k \in C_b(E), \quad \lim_{N\to\infty} \langle v_N, \phi_1 \otimes \cdots \otimes \phi_k \otimes 1 \cdots \otimes 1\rangle = \prod_i^k \langle v_1 \phi_i\rangle.$$

(3.1) can be seen to be equivalent to the fact that (see GRUNBAUM [1], SZNITMAN [1])

(3.2) $$\overline{X}_N = \frac{1}{N}\sum_1^N \varepsilon_{X^i} \quad (M(E) \text{ valued variables on } (E_N, v_N))$$

converge in law to the constant v.

So one considers as a realization of (1.1) the solution of

(3.3) $$X^i_t = X^i_0 + B^i_t + \frac{c}{N}\sum_{j\neq 1} L^\circ(X^i - X^j), \quad 1 \leq i \leq N,$$

where $$X^i_0 \notin S = \bigcup_{\substack{i,j,k \\ \text{distinct}}} \{x_i = x_j = x_k\}.$$

Theorem 3.1: If the initial law u_N of (X^i_0) put no mass on S and are u-chaotic, the law P_N on $C(R_+,R)^N$ of the process X_i are P-chaotic where P is the law of a process whose time marginal evolve according to Burgers' equation with initial condition u_0.

This process corresponding to P, is obtained formally by replacing the summation sign in (3.3) by an integration over an independent copy of the process, so that one considers

$$(3.4)\qquad \begin{aligned} &X_t = X_0 + B_t + c\, E_Y[L^\circ(X - Y)_t], \quad Y \text{ independent copy of } X,\\ &\text{law of } X_0 \text{ is } u_0. \end{aligned}$$

(3.4) is "well posed" as the following result shows:

Theorem 3.2: One has existence and uniqueness in law for the solutions of (3.4).

Idea of the proof:

The existence part follows in fact from the proof of Theorem 3.1. For the uniqueness part, one can use BARLOW-YOR's estimates [1] on local time and prove that X_t has a density $u(t,x)$ in $L^2([0,T] \times R)$, so that in fact $X_t = X_0 + B_t + 2c \int_0^t u(s,X_s)ds$, and $u(s,x)dx$ is the law of X_s.

One sees that $u(t,x)$ is a solution in the distribution sense of (1.2). Using now the Cole-Hopf transform, one defines

$$(3.5)\qquad W_t = \exp - 4c \int_{-\infty}^{x} u_t(dy),$$

and one checks using regularization that W_t satisfies in the distribution sense the heat equation, from this one gets the fact that $W_.$ is smooth and

$$(3.6)\qquad W_t = W_0 * \rho_t, \quad \text{with} \quad \rho_t(x) = \frac{1}{\sqrt{2\pi t}} \exp - \frac{x^2}{2t}$$

and from this one obtains easily the uniqueness part. □

Let us now indicate the idea of the proof of the theorem 3.1: the first idea is to avoid any direct comparison of (3.3) with (3.4), but rather, having in mind the equivalent property (3.2), to the fact of being chaotic, to define the empirical distributions

$$(3.7)\qquad \overline{X}_N = \frac{1}{N(N-1)} \sum_{i \neq j} \varepsilon_{(X^i_.,B^i_.,X^j_.,B^j_.,L^\circ(X^i - X^j)_.)}$$

and to show that these empirical measures are going to concentrate on a probability Q, whose first marginal will be the law P (unique solution of (3.4)). The first step is

1) Tightness of the laws $\overline{X}_N$:

This part is delicate. It boils down to checking that the laws of $X^{1,N}_{\cdot}$, $L^\circ(X^1 - X^2)^N_{\cdot}$ are tight.

The first idea would be use Tanaka's formula for the local times (see YOR [1]), to develop inside (3.3). But this idea does not work for c big. Instead, one uses the increasing reordering, $Y^1_t \leq \dots \leq Y^N_t$ of the (X^i_t), which form an oblique reflecting Brownian motion in the convex set $\{x_1 \leq \dots \leq x_N\}$. One then compares this process to the normally reflected Brownian motion in this convex set (see TANAKA [1]) and this way, one derives the necessary estimates.

2) Identification of the limit point of the laws of $\overline{X}_N$.

Let $\overline{P}_\infty$ be a limit point of these laws (it is a probability on the set of probability laws of the continuous processes (X^1,B^1,X^2,B^2,A), the last component being increasing).

One sees easily that for P_∞ a.e. m, (X^1,B^1) and (X^2,B^2) are i.i.d., that (B^1,B^2) is a 2-dimensional Brownian motion for the natural filtration generated by $(X^1.B^1,X^2,B^2,A)$, X^1_0 is u-distributed (as X^2_0), and

$$X^1_t = X^1_0 + B^1_t + A^1_t, \text{ with } A^1_t = c\, E_m[A_t/(X^1,B^1)].$$

The difficult point is to show that for $\overline{P}_\infty$ a.e. m, $A_t = L^\circ(X^1 - X^2)_t m$ a.s. (from that one infers that under m, X^1 is a solution of (3.4) and consequently has a completely determined law).

In order to show that $A_t = L^\circ(X^1 - X^2)_t$, one considers

$$(3.8)\quad H_t = |X^1_t - X^2_t| - |X^1_0 - X^2_0| - \int_0^t \operatorname{sign}(X^1_s - X^2_s)dA^1_s - \int_0^t \operatorname{sign}(X^2_s - X^1_s)dA^2_s - A_t.$$

If we can prove that H_t is a martingale under m (for $\overline{P}_\infty$ a.e. m), by Tanaka' formula for local times, we will see that $A_t = L^\circ(X^1 - X^2)_t$. Using ideas of lower semicontinuity and upper semicontinuity, one sees that

$$(3.9)\quad dL^g(X^1 - X^2)_t - 2.1(X^1_t = X^2_t)dA^2_t \leq dA_t \leq dL^g(X^1 - X^2)_t + 2.1(X^1_t = X^2_t)dA^1_t,$$

t

(here $L^g(X^1 - X^2)_t = \lim_{x \to 0^-} L^x(X^1 - X^2)_t$).

Projecting the right inequality of (3.9) on (X^1,B^1) and using the independence under m we have

$$\frac{1}{c}\, dA_t^1 \leq d\, E_{(X^2,B^2)}[L^g(X^1 - X^2)_t] + 2p(t,X_t^1)dA_t^1, \tag{3.10}$$

where $p(s,x) = m(X_s^1 = x)$.

One looks at the "bad" set $F = \{t, \exists\, x \in R,\ p(t,x) \geq 1/8c\}$.

It is a closed set of zero Lebesgue measure (one has the existence of an $L^2([0,T] \times R)$ density for the law of X_s under m). On the complement of this set F, $dA^1 << ds$, (by the use of (3.10) and an argument as in the proof of theorem 3.2). From this one obtains that $dA_t = dL^g(X^1 - X^2)_t = dL^o(X^1 - X^2)_t$ on F^c. So if I is an open interval in F^c, X_t^1, $t \in I$, is seen to be a solution of the non linear problem (3.4), so that

$$\exp - 4c\, F_t(x) = \exp - 4c\, F_s * \rho_{t-s}(x) \quad \text{if} \quad F_t(x) = m(X_t \leq x),\ s < t \ \text{ in } I.$$

This prevents the possibility of singularities at strictly positive times, so that $F \subset \{0\}$, and $X^1_{\cdot}$ is P-distributed, (one can also see that $\overline{P}_\infty$ is supported by a unique probability Q). With the help of (3.2), one obtains this way theorem 3.1.

□

Remark: In the case where $u_0(dx) = u_0(x)dx$, $u_0(\cdot)$ bounded measurable, one has trajectorial existence and uniqueness in (3.3), and if $u_N = u^{\otimes N}$, in that case if $\overline{X}^i$ denotes the solution of (3.4) constructed on X_0^i and B^i, one has the trajectorial convergence result: $\sup_{s \leq t} |X_s^{i,N} - \overline{X}_s^i| \to 0$ in probability.

REFERENCES

ARONSON, D.G. [1]: Bounds for the fundamental solution of a parabolic equation. Bull. Amer. Math. Soc. 73, 890-896, (1967).

BARLOW, M.T.; YOR, M. [1]: Semi-martingales inequalities via the Garsia-Rodemich-Rumsey lemma, and applications to the local times. J. Funct. Anal. 48, 2, 198-229, (1982).

CALDERONI, P., PULVIRENTI, M.[1]: Propagation of chaos for Burgers' equation. Ann. Inst. H. Poincaré, Sect. A (N.S), 39,1,85-97, (1983).

GRUNBAUM, F.A. [1]: Propagation of chaos for the Boltzmann equation. Arch. Rat. Mech. Anal. 42, 323-345, (1971).

GUTKIN, E., KAC, M. [1]: Propation of chaos and the Burgers' equation. SIAM J. Appl. Math. 43, 971-980, (1983).

KOTANI, S., OSADA, H. [1]: Propagation of chaos for Burgers' equation. J. Math. Soc. Japan, 275-294, (1985).

McKEAN, H.P. [1]: Propagation of chaos for a class of non linear parabolic equations. Lecture series on differential equations, vol. 7, 41-57, Catholic University, Washington, D.C., (1967).

OELSCHLAGER [1]: A law of large number for moderately interacting diffusion processes. Z. Wahrscheinlichkeitstheorie verw. Gebiete 69, 279-322, (1985).

SZNITMAN, A.S., VARADHAN, S.R.S. [1]: A multidimensional equation involving local time, to appear in Z. Wahrscheinlichkeitstheorie.

SZNITMAN, A.S. [1]: Equations de type Boltzmann spatialement homogènes. Z. Wahrscheinlichkeitstheorie verw. Gebiete, 66, 559-592, (1984).

[2]: A propagation of chaos result for Burgers' equation. To appear in Z. Wahrscheinlichkeitstheorie.

TANAKA, H. [1]: Stochastic differential equations with reflecting boundary condition in convex regions. Hiroshima Math. J. 9, 163-177, (1979).

[2]: Limit theorems for certain diffusion processes with interaction. Taniguchi Symp. SA, Katata 1982, 469-488.

YOR, M. [1]: Sur la continuité des temps locaux associés à certaines semi-martingales. Temps locaux, Asterisque 52-53, 23-35, S.M.F., Paris, (1978).

LIMIT DISTRIBUTIONS FOR ONE-DIMENSIONAL DIFFUSION PROCESSES IN SELF-SIMILAR RANDOM ENVIRONMENTS

H. Tanaka

Department of Mathematics
Faculty of Science and Technology
Keio University, Yokohama, JAPAN

Introduction

Let $X(t)$ be the one-dimensional diffusion process described by the stochastic differential equation

$$dX(t) = dB(t) - \frac{1}{2} W'(X(t))dt, \qquad X(0) = 0, \tag{1}$$

where $B(t)$ is a one-dimensional Brownian motion starting at 0 and $\{W(x), x \in \mathbb{R}\}$ is a random environment which is independent of the Brownian motion $B(t)$. We are interested in the asymptotic behavior of $X(t)$ as $t \to \infty$: Under what scaling does $X(t)$ have a limit distribution? Similar problems for random walks were considered by Kesten, Kozlov and Spitzer [5] and Sinai [8]. The problem we discuss here is a diffusion analogue of Sinai's random walk problem [8]. In the case of a Brownian environment Brox [1] proved that the distribution of $(\log t)^{-2}X(t)$ is convergent as $t \to \infty$. Similar results were obtained by Schumacher [7] for a considerably wider class of self-similar random environments. As was seen by these works the assumption of the self-similarity of the random environment is important and the notion of suitably defined valleys of the environment plays a central role in the proof. However, it was assumed that the environment has only one point which gives the same value of local minima or maxima (the bottom of a valley consists of a single point), and the explicit form of the limit distribution was unknown until a recent discovery by Kesten ([6]) for Sinai's random walk which corresponds to the case of a Brownian environment in our diffusion setup (Golosov also obtained the same result as Kesten's; see also Golosov [2] for the corresponding result in another different model).

In this paper we discuss the following three typical examples of random environments with emphasis on finding the limit distributions:

(i) Nonpositive reflected Brownian environment.

(ii) Nonnegative reflected Brownian environment.

(iii) Symmetric stable environment.

In the first two examples the environment has (uncountably) many points giving the same value of local maxima or minima. The proof in (ii) is only sketched. For details see [9]. The result in (iii) on the limit distribution is an extension of Kesten's result [6].

In [4] a unified definition of valleys of an environment is given in a general setup containing the above examples and some results similar to Brox's and Schumacher's are obtained, but here we limit ourselves to the above examples because we are interested mainly in the form of the limit distributions and we have explicit results concerning this only in some special cases.

The author wishes to thank K. Kawazu and Y. Tamura; his frequent discussions with them were very valuable.

1. Preliminaries and the Result of Brox

1.1. Given a real valued right continuous function $W(x)$ defined on the real line $\mathbb{R}$ and having left limits, we consider the stochastic differential equation (1) with environment $W(x)$. Since $W(x)$ is not differentiable in general, what is meant by a solution of (1) will need explanation. However, without considering a solution itself for a given Brownian motion $B(t)$, we just interprete the diffusion described by (1) as a diffusion process starting at 0 with generator

$$\frac{1}{2} e^{W(x)} \frac{d}{dx} \left(e^{-W(x)} \frac{d}{dx}\right). \tag{1.1}$$

Such a diffusion can be obtained from a Brownian motion through a scale change and a time change ([3]). To be precise let

$$\Omega = C([0,\infty) \to \mathbb{R})^{1)} \cap \{\omega(0) = 0\},$$

$$P = \text{the Wiener measure on } \Omega,$$

$$B(t) = \omega(t) = \text{the value of } \omega \text{ at time } t,$$

$$L(t,x) = \lim_{\varepsilon \downarrow 0} \frac{1}{\varepsilon} \int_0^t 1_{[x,x+\varepsilon)}(B(s))ds \qquad \text{(local time)},$$

$$S(x) = \int_0^x e^{W(y)}dy ,$$

$$A(t) = \int_0^t e^{-2W(S^{-1}(B(s)))}ds = \int_{\mathbb{R}} e^{-2W(S^{-1}(x))}L(t,x)dx, \ t > 0,$$

S^{-1}, A^{-1} = the inverse functions.

Then the process $X(t,W) = S^{-1}(B(A^{-1}(t))$ defined on the probability space (Ω,P) is a diffusion process with generator (1.1) starting at 0. The Brownian motion $B(t)$ used here is not the same as the one in (1) but we use the same notation. Let $(W^x)(\cdot) = W(\cdot+x)$. For a fixed $x \in \mathbb{R}$ we replace W in $X(t,W)$ by W^x and then consider

$$X^x(t,W) = x + X(t,W^x).$$

Then $X^x(t,W)$ is a diffusion process with generator (1.1) starting at x. In this paper we shall deal with $X(t) = X(t,W)$ and $X^x(t) = X^x(t,W)$ and the following notation will be used throughout.

$$S_\lambda(x) = \int_0^x e^{\lambda W(u)}du,$$

W_λ: a function (environment) defined by $W_\lambda(x) = \lambda^{-1}W(\lambda^2 x)$ for $\forall\, x \in \mathbb{R}$, $\lambda > 0$ being fixed.

In the following lemma due to Brox the medium W is fixed. For the proof see [1].

Lemma 1.1 ([1]). For each $\lambda > 0$

$$\{X(t,\lambda W_\lambda),\ t \geq 0\} \overset{d}{=} \{\lambda^{-2}X(\lambda^4 t,\ W),\ t \geq 0\}$$

where $\overset{d}{=}$ means the equality in distribution.

---

1) For a topological space R the notation $C([0,\infty) \to R)$ stands for the space of R-valued continuous functions defined on $[0,\infty)$.

When we consider the environment to be random, we denote by Q the probability distribution (on the space of environments) of the random environment. Since we are assuming that the random environment and the Brownian motion $B(t)$ are independent, the full distribution is $\mathcal{P} = P \times Q$. Thus, when the environment is fixed $X(t)$ is governed by P, and when the environment is random $X(t)$ is governed by $\mathcal{P}$.

1.2. In this subsection we limit ourselves to the case of continuous environment and state a result of Brox [1] in a form which is convenient for our use in §2.

Given a continuous function W on $\mathbb{R}$ which is supposed to be the environment in (1), we define a valley of W following [1]. For $x \neq y$ we put

$$H_{x,y} = \sup \{W(y') - W(x')\} \tag{1.2}$$

where the supremum is taken over all pairs of x' and y' such that $x < x' \leq y' < y$ or $y < y' \leq x' < x$ according as $x < y$ or $y < x$. A triple $V = (a,b,c)$ is called a <u>valley</u> of W if

(i) $a < b < c$,

(ii) $W(b) < W(x) < W(a)$ for every $x \in (a,b)$,

$W(b) < W(x) < W(c)$ for every $x \in (b,c)$,

(iii) $H_{a,b} < W(c) - W(b)$, $H_{c,b} < W(a) - W(b)$.

For a valley $V = (a,b,c)$, $D = (W(a) - W(b)) \wedge (W(c) - W(b))$ and $A = H_{a,b} \vee H_{c,b}$ are called the <u>depth</u> of V and the <u>inner directed ascent</u> of V, where $u \wedge v$ (resp. $u \vee v$) denotes min $\{u,v\}$ (resp. max $\{u,v\}$). It is obvious that $A < D$.

<u>Theorem 1.1 (Brox [1])</u>. (i) Let $V = (a,b,c)$ be a valley of W with the depth D and the inner directed ascent A.

(i) Let T_λ^x be the exit time of (a,c) for the process $X_\lambda^x(t) = x + X(t,\lambda W^x)$. Then for any $\delta > 0$ and any closed interval $I \subset (a,c)$

$$\lim_{\lambda\to\infty} \inf_{x \in I} P\{e^{\lambda(D-\delta)} < T_\lambda^x < e^{\lambda(D+\delta)}\} = 1.$$

(ii) For any $\varepsilon>0$, any closed interval $I \subset (a,c)$ and for any closed interval $J \subset (A,D)$

$$\lim_{\lambda\to\infty} \sup_{\substack{x\in I\\ r\in J}} P\{|X^x(e^{\lambda r}, \lambda W) - b| > \varepsilon\} = 0.$$

First proving the above result and then making use of the scaling relation (Lemma 1.1), Brox derived his main theorem:

Theorem 1.2 (Brox [1]). For any $\varepsilon>0$

$$P\{|\lambda^{-2}X(e^{-\lambda}, W) - b_\lambda| > \varepsilon\} \to 0, \quad \lambda\to\infty$$

in probability with respect to the probability measure Q of the Brownian environment, where b_λ is the unique bottom of a valley $V_\lambda = (a_\lambda, b_\lambda, c_\lambda)$ of W_λ such that

(1.3) $$a_\lambda < 0 < c_\lambda, \quad A_\lambda < 1 < D_\lambda.$$

2. Nonpositive Reflected Brownian Environment

As a simple example in which the environment has (uncountably) many points giving the same value of local maxima we consider the case of a nonpositive reflected Brownian environment. Main ideas of this section grew out from the conversation with Y. Tamura.

We first introduce the space $\underline{\underline{W}} = C(\mathbb{R}\to(-\infty, 0]) \cap \{W(0) = 0\}$ and denote by Q the probability measure on $\underline{\underline{W}}$ with respect to which $\{W(u) : u\geq 0\}$ and $\{W(-u) : u\geq 0\}$ are independent reflected Brownian motions on $(-\infty,0]$. An essential difference between the present case and Brox's one is that there is no valley of W_λ satisfying (1.3) in the present case. For $W\in \underline{\underline{W}}$ we put

$$W^*(u) = W(u) - \inf_{[0,u]} W, \ u \geq 0 ,$$
$$= W(u) - \inf_{[u,0]} W, \ u \leq 0 ,$$

and define $a_\pm = a_\pm(W)$, $b_\pm = b_\pm(W)$ and $c_\pm = c_\pm(W)$ as follows.

1. $d_\pm = \pm \min \{u > 0 : W^*(\pm u) = 1\}$,

2. $b_\pm = \pm \min \{u > 0 : W(\pm u) = M_\pm\}$

   where $M_+ = \min_{[0,d_+]} W$ and $M_- = \min_{[d_-,0]} W$,

3. $c_- = \min \{u > b_- : W(u) = 0\}$,

   $a_+ = \max \{u < b_+ : W(u) = 0\}$,

4. $e_\pm = \pm \min \{u > d_\pm : W(\pm u) = W(b_\pm)\}$,

5. $a_- = \max \{u < b_- : W(u) = \max_{[e_-,b_-]} W\}$,

   $c_+ = \min \{u > b_+ : W(u) = \max_{[b_+,e_+]} W\}$.

Then $V_\pm = (a_\pm , b_\pm , c_\pm)$ are valleys of W with $a_- < b_- < c_- < 0 < a_+ < b_+ < c_+$, Q-a.s. As for the depth and the inner directed ascent we have $A_\pm < 1 < D_\pm$, Q-a.s., because $W(a_-) > W(d_-)$ and $W(c_+) > W(d_+)$, Q-a.s. Substracting a suitable null set from $\underline{\underline{W}}$ we may assume that the statements hold for all W in the subtracted space and so we often omit the phrase "Q-a.s."

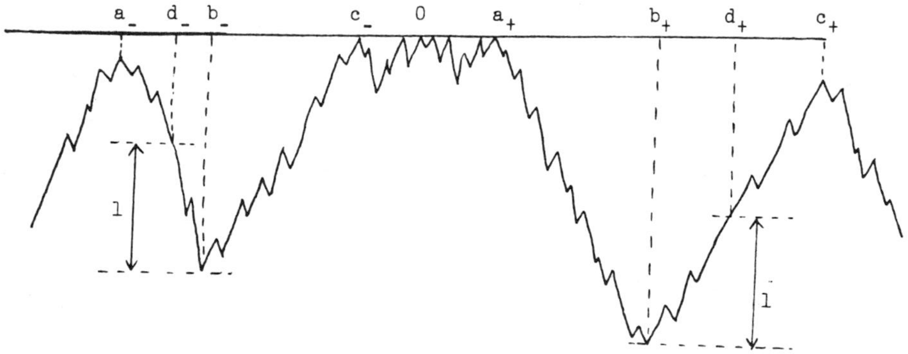

Our aim is to know the asymptotic behavior of $X(t,W)$ as $t\to\infty$, or more precisely, to find the limit distribution of $(\log t)^{-2}X(t,W)$ as $t\to\infty$. First we observe the process $X(t,\lambda W)$. If λ becomes large and time goes on, the process eventually falls into one of the valleys $V_\pm$ and thereafter Theorem 1.1 will tell about the asymptotic behavior of $X(e^{\lambda r}, \lambda W)$, $\lambda\to\infty$, $r\sim 1$. Then a use of the scaling relation (Lemma 1.1) will give the result for the limit distribution of $\lambda^{-2}X(e^{\lambda},W)$ as $\lambda\to\infty$.

To be precise choose α such that

(2.1) $$-\min_{[c_-,a_+]} W < \alpha < 1$$

and put

$$x_\pm = \pm \min\{u>0:\ W(\pm u) = -\alpha\},$$
$$T_\lambda^\pm = \min\{t \geq 0:\ X(t,\lambda W) = x_\pm\},\ \hat{T}_\lambda = T_\lambda^- \quad T_\lambda^+ .$$

Denote by $L_-(I,\xi)$ the local time at ξ for the reflected Brownian environment $\{W(u),\ u\in\mathbb{R}\}$, i.e.,

$$L_-(I,\xi) = \lim_{\varepsilon\downarrow 0}\frac{1}{\varepsilon}\int_I 1_{(\xi-\varepsilon,\xi]}(W(u))du,\ I = \text{interval in } \mathbb{R}.$$

<u>Lemma</u> <u>2.1</u>. (i) There exists $\delta = \delta(W) > 0$ such that

$$\lim_{\lambda\to\infty} P\{\hat{T}_\lambda < e^{\lambda(1-\delta)}\} = 1.$$

(ii) $$\lim_{\lambda\to\infty} P\{T_\lambda^- < T_\lambda^+\} = L_-([0,a_+],0)/L_-([c_-,a_+],0).$$

Proof. The condition (2.1) implies the existence of a modification $W^\#$ of W with the following properties (2.2) and (2.3).

(2.2) $W^\#\in C(\mathbb{R}\to\mathbb{R})$ and $W^\#(x) = W(x)$ for $\forall x\in[x_-,x_+]$.

(2.3) There exists a valley $V^\# = (a^\#, b^\#, c^\#)$ of $W^\#$ with depth $D^\# < 1$ and $a^\# < x_- < x_+ < c^\#$.

Let $T_\lambda^\# = \min\{t \geq 0:\ X(t,\lambda W^\#)\notin (x_-, x_+)\}$. Then for any $\delta>0$ such that $D^\# + 2\delta < 1$ we have

$$P\{\hat{T}_\lambda < e^{\lambda(1-\delta)}\} = P\{T^{\#}_\lambda < e^{\lambda(1-\delta)}\} \geq P\{T^{\#}_\lambda < e^{\lambda(D^{\#}+\delta)}\} \to 1, \ \lambda\to\infty$$

by (i) of Theorem 1.1. The proof of (ii) is easy; in fact

$$P\{T^-_\lambda < T^+_\lambda\} = \frac{S_\lambda(x_+) - S_\lambda(0)}{S_\lambda(x_+) - S_\lambda(x_-)}$$

$$= \frac{\int_{-\infty}^{0} e^{\lambda\xi}\, L_-([0,\ x_-],\ \xi)d\xi}{\int_{-\infty}^{0} e^{\lambda\xi}\, L_-([x_-,\ x_+],\ \xi)d\xi}$$

$$= \frac{\int_{-\infty}^{0} e^{\xi}\, L_-([0,x_+],\ \lambda^{-1}\xi)d\xi}{\int_{-\infty}^{0} e^{\xi}\, L_-([x_-,x_+],\ \lambda^{-1}\xi)d\xi}$$

$$\to L_-([0,x_+],\ 0)/L_-([x_-,x_+],\ 0),\ \lambda \to \infty$$

$$= L_-([0,a_+],\ 0)/L_-([c_-,\ a_+],\ 0).$$

<u>Lemma</u> <u>2.2</u>. There exist r_1 and r_2 with $r_1 < 1 < r_2$ such that for any $\varepsilon > 0$ and $r \in [r_1,r_2]$

$$\lim_{\lambda\to\infty} P\{X(e^{\lambda r},\ \lambda W) \in U_\varepsilon(b)\} = p$$

holds with $b = b_\pm$ and $p = p_\pm$ where $p_+ = L_-([c_-,0],0)/L_-([c_-,a_+],0)$, $p_- = 1 - p_+$ and $U_\varepsilon(b)$ denotes the ε-neighborhood of b. The above convergence is uniform with respect to r on $[r_1,\ r_2]$.

Proof. It is easy to see that there exist valleys $\tilde{V}_\pm = (\tilde{a}_\pm\ ,\ \tilde{b}_\pm\ ,\ \tilde{c}_\pm)$ of W with depth $\tilde{D}_\pm > 1$ and satisfying $a_- < \tilde{a}_- < b_- < x_- < \tilde{c}_- < c_-$ and $a_+ < \tilde{a}_+ < x_+ < b_+ < \tilde{c}_+ < c_+$, respectively. Denote by $\tilde{T}^{x+}_\lambda$ the exit time of $\tilde{I}_+ = (\tilde{a}_+\ ,\ \tilde{c}_+)$ for the process $X^{x_+}(t,\ \lambda W)$ and by $\tilde{T}^{x-}_\lambda$ the exit time of $\tilde{I}_- = (\tilde{a}_-,\ \tilde{c}_-)$ for the process $X^{x_-}(t,\lambda W)$. Then using the strong Markov property of $X(t,\lambda W)$ we have

$$P\{X(e^{\lambda(1-\delta)},\ \lambda W) \in \tilde{I}_+\}$$

(2.2)
$$\geq P\{T^+_\lambda = \hat{T}_\lambda < e^{\lambda(1-\delta)}\}\ P\{\tilde{T}^{x+}_\lambda > e^{\lambda(1-\delta)}\}$$

and a similar inequality with + replaced by - . We now take $\delta = \delta(W)$ of Lemma 2.1 and then let $\lambda\uparrow\infty$ in (2.2). Then the assertion (i) of Theorem 1.1 combined with Lemma 2.1 implies

$$\underline{\lim}_{\lambda\to\infty} P\{X(e^{\lambda(1-\delta)}, \lambda W) \in \Upsilon_+\} \geq P_+ \tag{2.3a}$$

and a similar inequality (= (2.3b)) with + replaced by - . (2.3a) and (2.3b) clearly imply that

$$\lim_{\lambda\to\infty} P\{X(e^{\lambda(1-\delta)}, \lambda W) \in \Upsilon_\pm\} = p_\pm . \tag{2.4}$$

We now take r_1 and r_2 such that

$$A_- \vee A_+ \vee (1-\delta) < r_1 < 1 < r_2 < D_- \wedge D_+$$

and prove the lemma with these r_1 and r_2. Let $r_1 \leq r \leq r_2$ and $t = e^{\lambda r} - e^{\lambda(1-\delta)}$. Then we have

$$\begin{aligned} &P\{X(e^{\lambda r}, \lambda W) \in U_\varepsilon(b_+)\} \\ &\quad = \int_{\Upsilon_+} P\{X(e^{\lambda(1-\delta)}, \lambda W) \in dx\} P\{X^x(t, \lambda W) \in U_\varepsilon(b_+)\} + o(1) \end{aligned}$$

which tends to p_+ uniformly in r as $\lambda\to\infty$ by Theorem 1.1 as applied to the valley V_+ because t can be expressed as $t = e^{\lambda r'}$ with $r'\to r$ as $\lambda\to\infty$. The other case $(b = b_-, p = p_-)$ can be proved similarly and so the proof is finished.

Since $b_\pm$ and $p_\pm$ are Borel functions of W we can define a probability measure μ_- on R by

$$\int \phi(x)\mu_-(dx) = \int\{p_- \phi(b_-) + p_+\phi(b_+)\}Q(dW), \quad \phi \in C_b(\mathbb{R}).$$

<u>Theorem 2.1</u>. The full distribution of $(\log t)^{-2}X(t)$ converges to μ_- as $t \to \infty$.

Proof. Lemma 2.2 implies that the full distribution of $X(e^{\lambda r(\lambda)}, \lambda W)$ converges to μ_- as $\lambda\to\infty$ provided that $r(\lambda)\to 1$ as $\lambda\to\infty$. Since our reflected Brownian environment is self-similar, i.e., $\{W_\lambda(x), x \in \mathbb{R}\}$ is again a reflected

Brownian environment, we see that the full distribution of $X(e^{\lambda r(\lambda)}, \lambda W_\lambda)$ also converges to μ_-. Applying the scaling relation (Lemma 1.1) we see that the full distribution of $\lambda^{-2}X(\lambda^4 e^{\lambda r(\lambda)}, W)$ converges to μ_-. If we take $r(\lambda) = 1 - 4\lambda^{-1}\log\lambda$, the last statement is nothing but the assertion of the theorem.

Theorem 2.2. μ_- has a symmetric density and

$$(2.5) \qquad \int_0^\infty e^{-\lambda x}\mu_-(dx) = \int_0^\infty \frac{d\sigma}{(\sigma+1)^2(\sigma + \sqrt{2\lambda}\coth\sqrt{2\lambda})\cosh\sqrt{2\lambda}}, \quad \lambda > 0.$$

Proof. If $d_\varepsilon(t)$ denotes the number of times that the reflected Brownian path $\{W(u) : u \geq 0\}$ crosses down from 0 to $-\varepsilon$ before t, then $\lim_{\varepsilon\downarrow 0} 2\varepsilon d_\varepsilon(t) = L_-([0,t],0)$, $t \geq 0$, Q-a.s. (see [3]). Therefore, putting $f_+ = \min\{u>0 : W(u) = -1\}$ and $L_2 = L([0,a_+],0)$ we can write for $\sigma, \lambda > 0$

$$(2.6) \qquad E^Q\{e^{-\sigma L_2 - \lambda f_+}\}$$

$$= \lim_{\varepsilon\downarrow 0}\sum_{n=0} e^{-\sigma(2\varepsilon n)} E_0^Q\{e^{-\lambda H_{-\varepsilon}}\}^n E_{-\varepsilon}^Q\{e^{-\lambda H_0}; H_0 < H_{-1}\}^{n-1}$$

$$\cdot E_{-\varepsilon}^Q\{e^{-\lambda H_{-1}}; H_{-1} < H_0\},$$

where H_a denotes the hitting time to a for the reflected Brownian motion on $(-\infty,0]$ and the suffix x in E_x^Q indicates that the initial position is x. Using the explicit form of $E^Q\{e^{-\lambda H_{-\varepsilon}}\}$, etc., the right hand side of (2.6) can be computed. The result is

$$(2.7) \qquad E^Q\{e^{-\sigma L_2 - \lambda f_+}\} = \frac{2\sqrt{2\lambda}}{e^{\sqrt{2\lambda}} - e^{-\sqrt{2\lambda}}} \cdot \frac{1}{2\sigma + c(\lambda)},$$

where $c(\lambda) = (e^{\sqrt{2\lambda}} + e^{-\sqrt{2\lambda}})(e^{\sqrt{2\lambda}} - e^{-\sqrt{2\lambda}})^{-1}\sqrt{2\lambda}$. In particular L_2 is exponentially distributed with mean 2. In a similar spirit we see easily that

$$(2.8) \qquad E^Q\{e^{-\lambda g_+}\} = \frac{e^{\sqrt{2\lambda}} - e^{-\sqrt{2\lambda}}}{\sqrt{2\lambda}(e^{\sqrt{2\lambda}} + e^{-\sqrt{2\lambda}})},$$

where $g_+ = b_+ - f_+$. Let $L_1 = L_-([c_-, 0], 0)$. Then

$$\int_0^\infty e^{-\lambda x}\ \mu_-(dx) = E^Q\{\frac{L_1}{L_1 + L_2} \cdot e^{-\lambda b_+}\}$$

$$= \int_0^\infty E^Q\ \{L_1 e^{-\sigma(L_1 + L_2) - \lambda f_+ - \lambda g_+}\} d\sigma ,$$

and making use of (2.7), (2.8) and the fact that L_1, g_+ and $\{L_2, f_+\}$ are independent we can compute the right hand side of the above. We thus obtain (2.5).

3. Nonnegative Reflected Brownian Environment

In this section we consider the case of a nonnegative reflected Brownian environment. This is a typical case where the bottom of a valley consists of (uncountably) many points.

Let $\underline{\underline{W}}$ be the space $C(\mathbb{R} \to [0,\infty)) \cap \{W(0) = 0\}$ and consider the probability measure Q on $\underline{\underline{W}}$ with respect to which $\{W(u):u \geq 0\}$ and $\{W(-u):u \geq 0\}$ are independent reflected Brownian motions on $[0,\infty)$. The study of asymptotic behaviors of $X(t,W)$ as $t \to \infty$ can be done by a method similar to that of Brox [1] as will be sketched here.

The definition of a valley given in 1.2 must be slightly modified. Given $W \in \underline{\underline{W}}$, a quartet $V = (a, b_1, b_2, c)$ is called a <u>valley</u> of W if

(i) $a < b_1 < 0 < b_2 < c$,

(ii) $W(b_1) = W(b_2) = 0$, $W(a) = W(b) = D > 0$,

(iii) $0 < W(x) < W(a)$ for $0 < x < b_1$,

$0 < W(x) < W(c)$ for $b_2 < x < c$,

(iv) $A = H_{a,b_2} \vee H_{c,b_1} < D$.

D and A are called the <u>depth</u> and the <u>inner directed ascent</u> of V, respectively. There exists a valley of W such that $A < 1 < D$ with Q-probability 1. In fact, let

$$a' = \max\{u < 0 : W(u) = 1\} , \quad c' = \min\{u > 0 : W(u) = 1\} ,$$

$$b_1 = \min\{u > a' : W(u) = 0\} , \quad b_2 = \max\{u < c' : W(u) = 0\} .$$

Then with a suitable choice of a and c with $a < a'$, $c' < c$, the quartet $V = (a, b_1, b_2, c)$ becomes a valley of W with $A < 1 < D$, Q-a.s. In what follows $V = (a, b_1, b_2, c)$ denotes such a valley of W. We first observe $X(t,\lambda W)$. As in [1] we can prove that $X(e^{\lambda r}, \lambda W)$ falls into an ε-neighborhood of $[b_1, b_2]$ as $\lambda \to \infty$, $r \sim 1$. Then how is $X(e^{\lambda r}, \lambda W)$ distributed on $[b_1, b_2]$ in the limit? This limit distribution can be identified with the limit, as $\lambda \to \infty$, of the invariant probability measure m_λ of the diffusion process on $[a,c]$ with (local) generator $\frac{1}{2} e^{\lambda W(x)} \frac{d}{dx} (e^{-\lambda W(x)} \frac{d}{dx})$ and with reflecting barriers at a and c. If $L_+(I, \xi)$ denotes the local time at ξ for the reflected Brownian environment $\{W(x), x \in \mathbb{R}\}$, then for an interval $I \subset [a,c]$

$$m_\lambda(I) = \int_I e^{-\lambda W(y)} dy \Big/ \int_a^c e^{-\lambda W(y)} dy$$

$$= \int_0^\infty e^{-\lambda\xi} L_+(I,\xi) d\xi \Big/ \int_0^\infty e^{-\lambda\xi} L_+([a,c], \xi) d\xi.$$

$$\to L_+(I \cap [b_1, b_2], 0)/L_+([b_1, b_2], 0).$$

Define a probability measure m_W in $\mathbb{R}$ by

$$m_W([u,v]) = L_+(I', 0)/L_+([b_1, b_2], 0)$$

where $I' = [u,v] \cap [b_1, b_2]$. Then for any interval I in $\mathbb{R}$ and for any family $\{r(\lambda), \lambda > 0\}$ satisfying $\lim_{\lambda\to\infty} r(\lambda) = 1$ we have

$$\lim_{\lambda\to\infty} P\{X(e^{\lambda r(\lambda)}, \lambda W) \in I\} = m_W(I)$$

for almost all W with respect to Q. Now we define $\mu_+ = \int m_W Q(dW)$. Then

$$\lim_{\lambda\to\infty} \mathcal{P}\{X(e^{\lambda r(\lambda)}, \lambda W) \in I\} = \mu_+(I).$$

Substituting W in the above by the scaled W_λ and then using Lemma 1.1 we obtain the following result.

<u>Theorem 3.1 ([9]).</u> (i) The full distribution of $(\log t)^{-2} X(t)$ converges to μ_+ as $t \to \infty$. (ii) μ_+ has a symmetric density and for $\lambda > 0$

$$\int_0^\infty e^{-\lambda x}\,\mu_+(dx) = \int_0^\infty \frac{\cosh\sqrt{2\lambda} - 1}{(\sigma+1)^3\{\cosh\sqrt{2\lambda} + \sigma \frac{\sinh\sqrt{2\lambda}}{\sqrt{2\lambda}}\}\lambda}\, d\sigma \,, \ \lambda > 0.$$

4. Symmetric Stable Environment

In this section the space $\underline{\underline{W}}$ of the environments is $D(\mathbb{R} \to \mathbb{R})$[1)] $\cap \{W(0) = 0\}$ and Q is the probability measure on $\underline{\underline{W}}$ with respect to which $\{W(u) : u\geq 0\}$ and $\{W(-u) : u\geq 0\}$ are independent and symmetric stable processes with exponent $\alpha(0 < \alpha < 2)$ such that

$$E^Q\{e^{\sqrt{-1}\xi W(u)}\} = e^{-|u||\tau|^\alpha} \,, \ u\in \mathbb{R}, \quad \xi \quad \mathbb{R}.$$

This case is contained in the frame of Schumacher's work [7] (see also Kawazu-Tamura-Tanaka [4] for additional information) and so the full distribution of $(\log t)^{-\alpha} X(t,W)$ is convergent as $t \to \infty$. The purpose this section is to give a simple probabilistic description of the limit distribution. In the case of a Brownian environment Kesten [6] proved that the density of the limit distribution is given by

$$(4.1) \qquad \frac{2}{\pi}\sum_{k=0}^{\infty}\frac{(-1)^k}{2k+1}\, e^{-(2k+1)^2\pi^2|x|/8} \,, \ x\in \mathbb{R}.$$

We begin by stating some known results (see [7], [4]). First we give the definition of a valley, which is a slight modification of the one given in §1. Let $W\in \underline{\underline{W}}$. By definition W is said to be <u>oscillating</u> at x $(\in \mathbb{R})$ if

$$\sup_{I_\pm} W > W(x\pm) \,, \ \inf_{I_\pm} W < W(x\pm)$$

for any $\varepsilon > 0$, where $I_+ = (x, x + \varepsilon)$ and $I_- = (x-\varepsilon, x)$. W is said to have a local minimum at x if $\inf_I W = W(x)\wedge W(x-)$ for some $\varepsilon>0$ where $I = (x-\varepsilon, x + \varepsilon)$. A local maximum is defined similarly. Denote by $\underline{\underline{W}}^\#$ the set of $W\in \underline{\underline{W}}$ with the following four properties.

---

1) This is the space of $\mathbb{R}$-valued right continuous functions on $\mathbb{R}$ with left limits.

(4.2a) $$\overline{\lim_{x\to\infty}} \{W(x) - \inf_{[0,x]} W\} = \overline{\lim_{x\to-\infty}} \{W(x) - \inf_{[x,0]} W\} = \infty .$$

(4.2b) If W is discontinuous at x, then W is oscillating at x.

(4.2c) For any open set G in $\mathbb{R}$ both the sets

$$\{x \in G : W(x) = \sup_G W\}, \ \{x \in G : W(x) = \inf_G W\}$$

contain at most one point.

(4.2d) W does not have a local maximum at $x = 0$.

Let $W \in \underline{\underline{W}}^{\#}$. Then $V = (a,b,c)$ is called a <u>valley</u> of W if

(i) $a < b < c$,

(ii) W is continuous at a, b and c,

(iii) $W(b) < W(x) < W(a)$ for every $x \in (a,b)$,

$W(b) < W(x) < W(c)$ for every $x \in (b,c)$,

(iv) $H_{a,b} < W(c) - W(b)$, $H_{c,b} < W(a) - W(b)$.

Here the notation $H_{x,y}$ is defined as in (1.2). The depth D and the inner directed ascent A are defined as in §1. Notice that (ii) implies that W is continuous at x if W has a local minimum (or maximum) at x. It can be proved that $Q(W^{\#}) = 1$. Moreover, making use of the self-similarity of a symmetric stable environment we can prove the following: There exists a valley $V = (a,b,c)$ of W such that $a < 0 < c$ and $A < 1 < D$, Q-a.s. The bottom of such a valley is uniquely determined by W and so is denoted by $b = b(W)$. For $\lambda > 0$ we define the scaled environment W_λ^α by $W_\lambda^\alpha(x) = \lambda^{-1}W(\lambda^\alpha x)$, $x \in \mathbb{R}$. Then as a special case of [6] (see also [4]) we have the following result: For any $\varepsilon>0$

$$P\{|\lambda^{-\alpha}X(e^\lambda, W) - b(W_\lambda^\alpha)| > \varepsilon\} \to 0$$

in probability with respect to Q as $\lambda\to\infty$ and consequently the full distribution of $(\log t)^{-\alpha}X(t)$ converges to the distribution of $b = b(W)$ as $t \to \infty$.

Our task is now to compute the distribution of $b = b(W)$. Although we are unable to give an analytic representation like (4.1), we have the following simple probabilistic representation of the limit distribution.

Theorem 4.1. The full distribution of $(\log t)^{-\alpha} X(t)$ converges, as $t \to \infty$, to the distribution with density

$$\phi(x) = \left\{ \frac{\Gamma(\alpha+1)}{\Gamma(\frac{\alpha}{2})} \right\}^2 \cdot Q\{ \sup_{[0,|x|]} W^* < 1\} , \quad x \in \mathbb{R},$$

where

$$W^*(u) = W(u) - \inf_{[0,u]} W, \quad u \geq 0.$$

The proof will be based on the following description of $b(W)$ due to Kesten. For $W \in \underline{\underline{W}}^{\#}$ we set

$$d_{\pm} = \pm \inf\{u > 0 : W^*(\pm u) \geq 1\}$$

where $W^*(-u)$, $u \geq 0$, is defined in a way similar to $W^*(u)$, and define $b_{\pm}$ by

$$W(b_+) = \inf_{[0,d_+]} W, \quad W(b_-) = \inf_{[d_-,0]} W.$$

We also set

$$V_{\pm} = W(b_{\pm}), \quad M_+ = \sup_{[0,b_+]} W, \quad M_- = \sup_{[b_-,0]} W.$$

Lemma 4.1 (Kesten [6]). Let (a, b, c) be a valley of W such that $a < 0 < c$ and $A < 1 < D$. Then

$$b = b_+ \text{ or } b_-$$

and the equality $b = b_+$ holds if and only if one of the following conditions holds:

(i) $V_- > V_+$ and $M_+ < (V_- + 1) \vee M_-$,

(ii) $V_- < V_+$ and $M_- > (V_+ + 1) \vee M_+$.

Since we deal only with the probability measure Q in the sequel, we write simply $E\{\cdot\}$ instead of $E^Q\{\cdot\}$ for the expectation with respect to Q. Again we introduce notation:

$$F_\lambda(x) = E\{e^{-\lambda b_+} ; V_+ < x - 1, M_+ < x\}$$

$$= E\{e^{-\lambda b_-} ; V_- < x - 1, M_- < x\} .$$

Clearly $F_\lambda(x)$ vanishes identically on $(-\infty, 0]$ and equals $E\{e^{-\lambda b_+}\}$ on $[1,\infty)$.

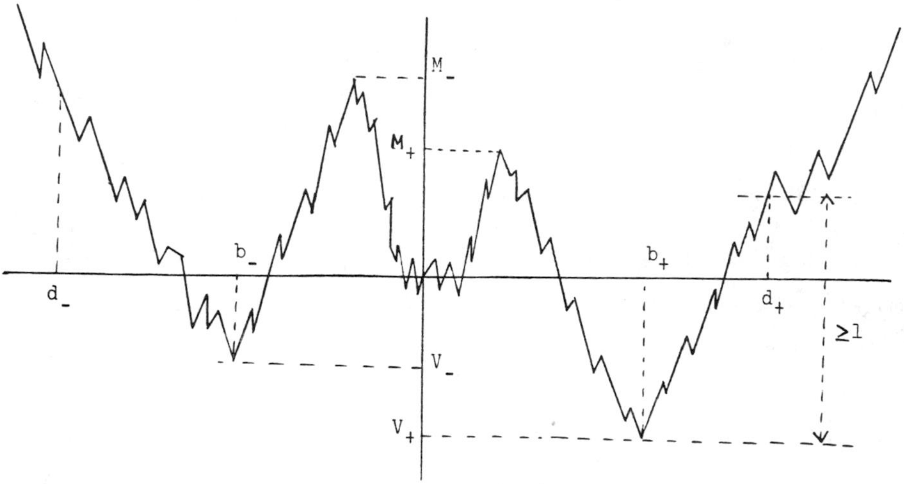

Lemma 4.2. $E\{e^{-\lambda b}; b > 0\} = \int_0^1 F_\lambda(x)dF_0(x)$, $\lambda > 0$.

Proof. Let

$$E_1 = \{V_- > V_+ , M_+ < (V_- + 1)\vee M_-\} ,$$

$$E_2 = \{V_1 < V_+ , M_- > (V_+ + 1)\vee M_+\} .$$

Then it is easy to see that

$$(E_1 \cup E_2)^c = \{V_- < V_+\vee(M_+ - 1), M_- < (V_+ + 1)\vee M_+\},$$

and hence

$$E\{e^{-\lambda b} ; b > 0\}$$

$$= E\{e^{-\lambda b_+} ; V_- > V_+ , M_+ < (V_- + 1)\vee M_-\}$$

$$+ E\{e^{-\lambda b_+} ; V_- < V_+ , M_- > (V_+ + 1)\vee M_+\}$$

$$= E\{e^{-\lambda b_+} ; E_1 \cup E_2\}$$

$$= E\{e^{-\lambda b_+}\} - E\{e^{-\lambda b_+}\ ;\ (E_1 \cup E_2)^c\}$$

$$= E\{e^{-\lambda b_+}\} - E\{e^{-\lambda b_+}\ ;\ V_- < V_+ \vee (M_+ - 1),\ M_- < (V_+ + 1) \vee M_+\}$$

$$= F_\lambda(1) - E\{e^{-\lambda b_+} \cdot F_0((V_+ + 1) \vee M_+)\}$$

$$= F_\lambda(1) - \int_0^1 F_0(x) dF_\lambda(x)$$

$$= \int_0^1 F_\lambda(x) dF_0(x).$$

Now the rest of the proof of Theorem 4.1 is divided into two parts.

1. Properties of a symmetric stable process. We deal with the symmetric stable process $\{x + W(t),\ t \geq 0,\ x \in \mathbb{R},\ Q\}$ and prepare some of its properties for our later use. For the proof omitted here, see [10]. For $x,\ a \in \mathbb{R}$ we set

$$S_a^x = \inf\{t \geq 0:\ x + W(t) \geq a\},$$

$$T_a^x = \inf\{t \geq 0 : x + W(t) \leq a\},$$

$$T^x = S_1^x \wedge T_0^x .$$

Let $g_\lambda(x,y)$ be the Green function of order $\lambda > 0$ of the symmetric stable process with absorbing barriers at 0 and 1. We thus have

$$E\{\int_0^{T^x} e^{-\lambda t} f(x + W(t)) dt\} = \int_0^1 g_\lambda(x,y) f(y) dy, \quad 0 \leq x \leq 1.$$

It is known that $g_\lambda(x,y)$ is symmetric in x and y. We still need the following notation:

$$p_\lambda^+(x) = E\{e^{-\lambda S_1^x}\ ;\ S_1^x < T_0^x\}\ ,\quad p_\lambda^-(x) = E\{e^{-\lambda T_0^x}\ ;\ T_0^x < S_1^x\}\ ,$$

$$p_\lambda(x) = p_\lambda^+(x) + p_\lambda^-(x) = E\{e^{-\lambda T^x}\}\ ,$$

$$q^+(x) = 2\alpha^{-1}\{\Gamma(\alpha/2)\}^{-2}\, x^{\frac{\alpha}{2}}\, (1-x)^{\frac{\alpha}{2}-1},$$

$$q^-(x) = 2\alpha^{-1}\{\Gamma(\alpha/2)\}^{-2}\, (1 - x)^{\frac{\alpha}{2}}\, x^{\frac{\alpha}{2}-1}\ ,$$

$$q_\lambda^{\pm}(x) = q^{\pm}(x) - \lambda \int_0^1 g_\lambda(x,y)q^{\pm}(y)dy,$$

$$r(x) = x^{\frac{\alpha}{2}-1}(1-x)^{\frac{\alpha}{2}-1}.$$

Notice that

$$g_\lambda(x,y) = g_\lambda(1-x,\ 1-y),$$

$$p_\lambda^{+}(x) = p_\lambda^{-}(1-x),\ p_\lambda(x) = p_\lambda(1-x),$$

$$q^{+}(x) = q^{-}(1-x)\ ,\ q_\lambda^{+}(x) = q_\lambda^{-}(1-x).$$

We also use the notation:

$$c(\alpha) = \Gamma(\alpha+1)\pi^{-1}\sin(\alpha\pi/2),$$

$$\langle p,\ q\rangle = \int_0^1 p(x)q(x)dx\ .$$

<u>Lemma</u> <u>4.3.</u> $\alpha\{\Gamma(\alpha/2)\}^2 \langle p_\lambda\ ,\ q^{+}\rangle = \langle p_\lambda\ ,\ r\rangle$.

Proof. The left hand side equals

$$2\langle p_\lambda,\ xr\rangle = 2\langle p_\lambda\ ,\ (1-x)r\rangle = \text{the right hand side};$$

the second equality is obtained by adding the preceding two terms and then dividing the sum by 2.

<u>Lemma</u> <u>4.4</u> (Watanabe [10]). (i) For any $f \in C([0,1])$

$$\lim_{x\downarrow 0} x^{-\alpha/2} \int_0^1 g_\lambda(x,y)f(y)dy = \langle f, q_\lambda^{-}\rangle,$$

$$\lim_{x\downarrow 0} x^{-\alpha/2} \int_0^1 g_\lambda(1-x,\ y)f(y)dy = \langle f,\ q_\lambda^{+}\rangle\ .$$

(ii) $$p_\lambda^{+}(x) = \alpha^{-1}c(\alpha) \int_0^1 g_\lambda(x,y)(1-y)^{-\alpha}\, dy,$$

$$p_\lambda^{-}(x) = \alpha^{-1}c(\alpha) \int_0^1 g_\lambda(x,y)y^{-\alpha}\, dy\ ,$$

$$p_0^{+}(x) = 2^{1-\alpha}\Gamma(\alpha)\{\Gamma(\alpha/2)\}^{-2} \int_{-1}^{-1+2x} (1-y^2)^{\frac{\alpha}{2}-1}\, dy\ .$$

Lemma 4.5. We have

(i) $\lim_{x \downarrow 0} p_\lambda^+(x)x^{-\alpha/2} = \lim_{x \downarrow 0} p_\lambda^-(1 - x)x^{-\alpha/2} = K - \lambda\langle p_\lambda^-, q^+\rangle$,

(ii) $\lim_{x \downarrow 0} \{1 - p_\lambda^-(x)\}x^{-\alpha/2} = \lim_{x \downarrow 0} \{1 - p_\lambda^+(1 - x)\}x^{-\alpha/2} = K + \lambda\langle p_\lambda^+, q^+\rangle$

where

$$K = \alpha^{-1}c(\alpha) \int_0^1 q^+(x)x^{-\alpha}dx = 2\Gamma(1 - \tfrac{\alpha}{2})c(\alpha)/\Gamma(\alpha/2).$$

Proof. We give the proof of (ii). Since

$$\int_0^1 g_\lambda(x,y)dy = \lambda^{-1}\{1 - p_\lambda(x)\}$$

we have

$$\lim_{x \downarrow 0} \{1 - p_\lambda^-(x)\}x^{-\alpha/2} = \lim_{x \downarrow 0} \{1 - p_\lambda(x)\}x^{-\alpha/2} + \lim_{x \downarrow 0} p_\lambda^+(x)x^{-\alpha/2}$$

$$= \lambda \lim_{x \downarrow 0} x^{-\alpha/2} \int_0^1 g_\lambda(x,y)dy + \lim_{x \downarrow 0} p_\lambda^+(x)x^{-\alpha/2}$$

$$= \langle\lambda + \alpha^{-1}c(\alpha) \cdot x^{-\alpha}, q_\lambda^+\rangle$$

$$= K + \lambda\langle p_\lambda, q^+\rangle - \lambda\langle p_\lambda^-, q^+\rangle$$

$$= K + \lambda\langle p_\lambda^+, q^+\rangle .$$

2. Computation of $E\{e^{-\lambda b}; b > 0\}$.

Lemma 4.6. For $\lambda, \mu > 0$

$$E\{e^{-\lambda b_+ - \mu(d_+ - b_+)}\} = \{K - \mu\langle p_\mu^-, q^+\rangle\}\{K + \lambda\langle p_\lambda^+, q^+\rangle\}^{-1}.$$

In particular

$$E\{e^{-\lambda b_+}\} = K\{K + \lambda\langle p_\lambda^+, q^+\rangle\}^{-1},$$

$$E\{e^{-\lambda d_+}\} = \{K - \lambda\langle p_\lambda^-, q^+\rangle\}\{K + \lambda\langle p_\lambda^+, q^+\rangle\}^{-1}.$$

Proof. The proof is similar to that in [6]. For $\varepsilon>0$ we set

$$T^{(0)} = T_\varepsilon^{(0)} = 0,$$

$$T^{(n)} = T_\varepsilon^{(n)} = \inf\{t > T^{(n-1)} : W(t) - W(T^{(n-1)}) < -\varepsilon\}$$

$$S^{(n)} = S_\varepsilon^{(n)} = \inf\{t > T^{(n-1)} : W(t) - W(T^{(n-1)}) > 1-\varepsilon\},\ n = 1,2,\ldots$$

Then

$$E\{e^{-\lambda b_+ - \mu(d_+ - b_+)}\}$$

$$= \lim_{\varepsilon\downarrow 0} \sum_{n=1}^{\infty} E\{e^{-\lambda T^{(n)} - \mu(S^{(n+1)} - T^{(n)})};\ T^{(k)} < S^{(k)} \text{ for } 1 \le \forall\ k \le n$$

$$\text{and } S^{(n+1)} < T^{(n+1)}\}$$

$$= \lim_{\varepsilon\downarrow 0} \sum_{n=1}^{\infty} \{E\{e^{-\lambda T^0_{-\varepsilon}};\ T^0_{-\varepsilon} < S^0_{1-\varepsilon}\}\}^n\, E\{e^{-\mu S^0_{1-\varepsilon}};\ S^0_{1-\varepsilon} < T^0_{-\varepsilon}\}$$

$$= \lim_{\varepsilon\downarrow 0} \sum_{n=1}^{\infty} \{p_\lambda^-(\varepsilon)\}^n \cdot p_\lambda^+(\varepsilon).$$

If we set

$$K_1 = K + \lambda\langle p_\lambda^+,\ q^+\rangle,\quad K_2 = K - \lambda\langle p_\lambda^-,\ q^+\rangle,$$

then

$$1 - p_\lambda^-(\varepsilon) \sim K_1\varepsilon^{\alpha/2},\quad p_\lambda^+(\varepsilon) \sim K_2\varepsilon^{\alpha/2} \quad \text{as} \quad \varepsilon\downarrow 0 .$$

Therefore we have

$$E\{e^{-\lambda b_+ - \mu(d_+ - b_-)}\} = \lim_{\varepsilon\downarrow 0} \sum_{n=1}^{\infty} (1 - K_1\varepsilon^{\alpha/2})^n K_2\varepsilon^{\alpha/2}$$

$$= K_2/K_1 = \{K - \lambda\langle p_\lambda^-,\ q^+\rangle\}\{K + \lambda\langle p_\lambda^+,\ q^+\rangle\}^{-1},$$

completing the proof of the lemma.

For the proof of the theorem it is enough to prove the following lemma.

<u>Lemma 4.7</u>. For $\lambda > 0$

$$E\{e^{-\lambda b};\ b>0\} = \{\Gamma(\alpha+1)/\Gamma(\alpha/2)\}^2 \int_0^\infty e^{-\lambda t} Q\{d_+ > t\}dt.$$

Proof. We first notice that

$$F_\lambda(x) = E\{e^{-\lambda b_+} ; V_+ < x - 1, M_+ < x\}$$

$$= E\{e^{-\lambda T^0_{x-1}} ; T^0_{x-1} < T^0_x\}E\{e^{-\lambda b_+}\} = p^+_\lambda(x)E\{e^{-\lambda b_+}\} ,$$

$$F_0(x) = p^+_0(x) = 2^{1-\alpha}\Gamma(\alpha)\{\Gamma(\alpha/2)\}^{-2} \int_{-1}^{-1+2x} (1 - y^2)^{\frac{\alpha}{2}-1} dy,$$

$$dF_0(x) = \Gamma(\alpha)\{\Gamma(\alpha/2)\}^{-2}r(x)dx .$$

We thus have

$$(4.3) \qquad E\{e^{-\lambda b} ; b > 0\} = \int_0^1 F_\lambda(x)dF_0(x)$$

$$= E\{e^{-\lambda b_+}\} \int_0^1 p^+_\lambda(x)\, dF_0(x) = \Gamma(\alpha)\{\Gamma(\alpha/2)\}^{-2} E\{e^{-\lambda b_+}\}\langle p^+_\lambda , r\rangle$$

$$= 2^{-1}\Gamma(\alpha)\{\Gamma(\alpha/2)\}^{-2}E\{e^{-\lambda b_+}\}\langle p_\lambda, r\rangle .$$

On the other hand, using Lemma 4.3 we have

$$\int_0^\infty e^{-\lambda t}Q\{d_+ > t\}dt = \lambda^{-1}\{1 - E(e^{-\lambda d_+})\} = \langle p_\lambda, q^+\rangle\{K + \lambda\langle p^+_\lambda, q^+\rangle\}^{-1}$$

$$= \alpha^{-1}\{\Gamma(\alpha/2)\}^{-2} \langle p_\lambda, r\rangle\{K + \lambda\langle p^+_\lambda, q^+\rangle\}^{-1}$$

$$= (\alpha K)^{-1}\{\Gamma(\alpha/2)\}^{-2} E\{e^{-\lambda b_+}\}\langle p_\lambda,r\rangle .$$

Combining this with (4.3) we obtain Lemma 4.7. This completes also the proof of Theorem 4.1.

Remark. The density ϕ in Theorem 4.1 is symmetric and

$$\int_0^\infty e^{-\lambda x}\phi(x)dx = \frac{\alpha\{\Gamma(\alpha)\}^2}{\{\Gamma(\alpha/2)\}^4} \cdot \frac{\langle p_\lambda , r\rangle}{K + \lambda\langle p^+_\lambda, q^+\rangle} , \lambda > 0.$$

References

[1] T. Brox, A one-dimensional diffusion process in a Wiener medium, to appear in Ann. Probab.

[2] A.O. Golosov, The limit distributions for random walks in random environments, Soviet Math. Dokl., 28 (1983), 19-22.

[3] K. Itô and H.P. McKean, Diffusion Processes and Their Sample Paths, Springer-Verlag, Berlin, 1965.

[4] K. Kawazu, Y. Tamura and H. Tanaka, One-dimensional diffusions and random walks in random media, in preparation.

[5] H. Kesten, M.V. Kozlov and F. Spitzer, A limit law for random walk in a random environment, Compositio Math., 30 (1975), 145-168.

[6] H. Kesten, The limit distribution of Sinai's random walk in random environment, to appear in Physica.

[7] S. Schumacher, Diffusions with random coefficients, Contemporary Mathematics (Particle Systems, Random Media and Large Deviations, ed. by R. Durrett), 41 (1985), 351-356.

[8] Y.G. Sinai, The limiting behavior of a one-dimensional random walk in a random medium, Theory of Probab. and its Appl., 27 (1982), 256-268.

[9] H. Tanaka, Limit distribution for 1-dimensional diffusion in a reflected Brownian medium, preprint, 1986.

[10] S. Watanabe, On stable processes with boundary conditions, J. Math. Soc. Japan, 14(1962), 170-198.